MATHEMATICS AND PROBLEM SOLVING FOR FIRE SERVICE PERSONNEL MANAGEMENT

Eugene F. Mahoney, MS, BS
Retired Battalion Chief, LAFD
Retired Fire Chief, Arcadia Fire Department
Retired Professor of Fire Science

Brady
is an imprint of
Pearson

Boston Columbus Indianapolis New York San Francisco Upper Saddle River Amsterdam
Cape Town Dubai London Madrid Milan Munich Paris Montreal Toronto Delhi
Mexico City Sao Paulo Sydney Hong Kong Seoul Singapore Taipei Tokyo

Publisher: Julie Levin Alexander
Editor-in-Chief: Marlene McHugh Pratt
Senior Acquisitions Editor: Stephen Smith
Senior Managing Editor for Development:
  Lois Berlowitz
Development Editor: Eileen M. Clawson
Associate Editor: Monica Moosang
Editorial Assistant: Heather Luciano
Director of Marketing: Karen Allman
Executive Marketing Manager: Katrin Beacom
Marketing Specialist: Michael Sirinides
Marketing Assistant: Judy Noh
Managing Production Editor: Patrick Walsh
Production Liaison: Frances Russello
Production Editor: Julie Li
Senior Media Editor: Amy Peltier
Media Project Manager: Lorena Cerisano
Creative Director: Jayne Conte
Cover Designer: Lisbeth Axell
Cover Image Credit: Courtesy of Rick McClure, LAFD
Compositor: Aptara®, Inc.
Printing and Binding: Edwards Brothers, Inc.

Many of the designations by manufacturers and sellers to distinguish their products are claimed as trademarks. Where those designations appear in this book, and the publisher was aware of a trademark claim, the designations have been printed in initial caps or all caps.

Copyright © 2010 by Pearson Education, Inc., Upper Saddle River, New Jersey 07458. Pearson Prentice Hall. All rights reserved. Printed in the United States of America. This publication is protected by Copyright and permission should be obtained from the publisher prior to any prohibited reproduction, storage in a retrieval system, or transmission in any form or by any means, electronic, mechanical, photocopying, recording, or likewise. For information regarding permission(s), write to: Rights and Permissions Department.

Library of Congress Cataloging-in-Publication Data
Mahoney, Gene
  Mathematics and problem solving for fire service personnel / Eugene Mahoney.
    p. cm.
  ISBN-13: 978-0-13-502825-4
  ISBN-10: 0-13-502825-6
  1. Mathematics—Problems, exercises, etc.   2. Problem solving—Problems, exercises, etc.
  3. Word problems—Problems, exercises, etc.   I. Title.
  QA43.M2285 2010
  510.76—dc22

                                                                2009014129

10 9 8 7 6 5 4 3 2 1

Brady
is an imprint of

www.bradybooks.com

ISBN-10:   0-13-502825-6
ISBN-13: 978-0-13-502825-4

This book is dedicated to my stepdaughter, Jeri Stapp, who devoted so many hours to mastering the principles established herein.

Contents

Preface   *xiii*
Acknowledgments   *xv*
About the Author   *xvii*
Introduction   *xix*

Chapter 1 Introduction to the World of Numbers   1

Suggestions for Readers	2
The Basic Concept of Numbers	2
Introduction to the Number System	3
An Explanation of Numbers	4
Determining Prime Numbers	5
Divisibility Rules	5
Determining Whether a Number Larger than 100 is a Prime Number	6
Positive and Negative Integers	7
Word Problems	8
Changing Word Problems to Number Problems	9
Sample Word Problems	9

Chapter 2 Solving and Proving Whole Number Addition Problems   13

A Review of the Addition Process	14
Proving Addition Problems	17
Clues to Errors	20
Proving Smaller Addition Problems	21
Introduction to Positive and Negative Integers	22
Adding Positive and Negative Integers	23
Practice Adding Positive and Negative Problems	24
Sample Word Problems	25

Chapter 3 Working with Whole Number Multiplication Problems — 38

A Review of the Multiplication Process	39
The Multiplication Tables	39
A Review of the Times Tables	39
Multiplying by Nine	40
Solving a Multiplication Problem	42
Proving a Multiplication Problem	45
Shortcut Multiplication Procedures	46
Multiplying a Two-Digit Number by Eleven	47
Mentally Multiplying Two 'Teen Numbers	47
Squaring a Number Ending in Five	49
Multiplying a Number by Ten	49
Multiplying a Number by 100	50
Multiplying Positive and Negative Integers	51

Chapter 4 Solving Whole Number Division, Subtraction, and Square Root Problems — 59

A Review of the Division Process	60
The Division Process	60
Determining Whether a Division Problem Has Been Solved Correctly	61
Dividing a Four-Digit Dividend by a Two-Digit Divisor	61
Determining Whether the Problem Has Been Solved Correctly	62
Determining Whether a Dividend Can Be Divided Evenly	62
Shortcuts for Division Problems	64
Dividing Positive and Negative Integers	65
A Review of the Subtraction Process	68
Proving a Subtraction Problem	69
Subtracting Negative and Positive Integers	70
A Review of the Square Root Process	77
Solving a Two-Digit Square Root Problem	80
Proving a Square Root Problem	82

Chapter 5 Working with Fractions — 87

Introduction to Fractions	88
Types of Common Fractions	88
Related Principles of Whole Numbers	89
Changing a Fraction to an Equivalent Fraction	90
Changing a Fraction to a Higher or Lower Term	90
Converting a Mixed Number to an Improper Fraction	91
Converting an Improper Fraction to a Mixed Number	91
Converting a Decimal to a Fraction	91
Converting Minutes to a Fraction	92

Addition of Mixed Numbers	98
Subtraction of Common Fractions	104
Subtracting Mixed Numbers	104
Multiplication of Common Fractions	109
The Multiplication Process	110
Canceling	110
Multiplying Whole Numbers and Common Fractions	112
Multiplying Mixed Numbers, Whole Numbers, and Common Fractions	113
Dividing a Whole Number by a Mixed Number	119
Dividing a Mixed Number by a Mixed Number	119

Chapter 6 Working with Decimals — 123

Rounding Numbers	125
Converting a Fraction to a Decimal	126
Converting a Decimal to a Fraction	126
Principles of Adding Decimals	128
Proving Decimal Addition Problems	129
Principles of Subtracting Decimals	134
Principles of Multiplying Decimals	138
The Principles of Dividing Decimals	145

Chapter 7 Solving Combination Problems — 151

Basic Sequence for Working Combination Problems	152
Important Information	157

Chapter 8 Understanding Measurements — 167

Length Measurements	169
A Word of Caution Regarding Metric Usage	170
Area Measurements	178
Pressure Measurements	182
Volume Measurements	191
Rectangular Containers	192
Cylindrical Containers	193
Weight Measurements	198
Determining the Weight of Water in a Container	199
Fluid Measurements	202
Temperature Measurements	207

Chapter 9 Working with Percent and Percentages — 211

Introduction to the Term Ratio	212
Introduction to Percent	212

Converting a Percent to a Decimal	212
Converting a Decimal to a Fraction	212
Converting a Percent to a Fraction	213
Converting a Decimal to a Percent	213
Converting a Fraction to a Percent	214
Solving Percent Problems	215
Solving Alternative Percent Problems	216
Introduction to Percentage	218
Introduction to Percentile	218
Tipping	218
Sale Items	219
Grocery Sale Items	220

Chapter 10 Working with Statistics — 224

Introduction to Statistics	225
The Standard Curve of Distribution	227
Introduction to Graphs and Charts	228
Types of Graphs and Charts	229
Development of a Graph or Chart	229
Information for the Reader	230

Chapter 11 Formulas and Formula Manipulations — 241

Solving Formula Problems	241
Manipulating Formulas	242
Working with Positive and Negative Units	242
Working with Multiplication Units	244
Working with Exponent Units	246
Working with Square Root Units	247
Double Trouble Situations	248

Appendix

Section One: U.S. Units of Measurement	*258*
Section Two: Metric Units of Measurement	*260*
Section Three: International Units of Measurement	*262*
Section Four: Addition Chart	*264*
Section Five: Multiplication Chart	*265*
Section Six: Times Tables	*266*
Section Seven: Placement Chart	*267*

Section Eight: Basic Fire Department Hydraulic Figures	*268*
Section Nine: Abbreviations	*269*
Section Ten: Traditional Formulas Introduced in the Book	*270*
Section Eleven: Square Roots	*274*
Reader's Practice Problems: Answers	*276*
Test Answers	*315*
Glossary	*326*
Index	*329*

Preface

Every day, firefighters depend upon their mathematical skills to do their job effectively. These skills are normally used to work hydraulic problems, both in the classroom and on the fire ground. However, working on fires is just one of the situations where a fire department may be dispatched today. Throughout the country, most engine companies are dispatched to more medical emergencies than to fires. Since 9/11, modern firefighters have become more involved in hazardous materials situations where they are called upon to use their math skills to calculate the area and volume of the hazardous material spills. As they advance up the ladder, they are continually called upon to use their mathematical skills in time management, budgetary considerations, fire company run statistics, and drug calculations when responding to emergency medical incidents. As part of the labor force, their union leaders use their math skills in an effort to compete with their counterparts in management.

When a coordinator of fire science programs at a community college looks at the magnitude of young men and women taking fire science courses or attending a fire academy to become a firefighter, his or her thoughts immediately wander to what lies ahead of the students if they are successful in their goal. Consequently, more and more of these coordinators and fire administrators see the need for a prospective firefighter to include the development of mathematical skills as a portion of their preservice training.

As a result, this book has been written as both a review of basic math for individuals who are studying to become firefighters, and a study book for those firefighters preparing for a promotional examination. Although traditional problems have been included to meet the needs of those requiring a basic review, mathematical word hydraulic problems have been included for both groups. The book has also been prepared with the realization that some readers will be using it in a formalized classroom situation whereas others will be using it for self-study. Consequently, a large number of problems have been included for the reader to solve—chapter problems, chapter review tests, and a final test for every chapter. Also included are Reader's Practice Problems scattered throughout the book. An explanation of how these practice problems are solved is included in the Appendix. When a reader is able to answer all the questions and work all of the problems without difficulty, he or she will have an abundant foundation for solving mathematical problems.

Acknowledgments

Over the years I have asked many students and scores of experienced firefighters what they thought was needed in a mathematics book for firefighters. The answers kept repeating themselves: lots of examples, abundant illustrations, and step-by-step solutions to problems. The format of this book reflects the requests of those people.

I have also sought the advice of many of my associates throughout my career, seeking opinions about course content and sequence of presentation. Their contributions to this book were invaluable.

Perhaps the greatest contributor to this manual is the Los Angeles City Fire Department (L.A.F.D.). It was the experience gathered there that provided me with the firefighting experience that is essential to the understanding of the fire department hydraulic principles. Like so many have said, it all comes together on the fire ground.

I wish to thank the following individuals who so willingly contributed to the production of this book. First is Rick McClure, L.A.F.D., who furnished all of the photos used in the book. Next is the staff at Pearson Education, in particular my editor, Stephen Smith, and associate editor, Monica Moosang. Special thanks is extended to developmental editor Eileen Clawson, who devoted so many hours to pulling together the numerous parts of the manuscript before submitting it to production.

I would also like to thank the following people for their help in reviewing this book:

Captain Pat Ballard
Dublin Fire Department
Dublin, Georgia

Rick Ducey
Community Emergency Response Team (CERT)
Montgomery County Fire & Rescue Service
Montgomery County, Maryland

Jeffrey Huber, MS, EFO
Professor of Fire Science
Lansing Community College
Lansing, Michigan

David J. Kirk
Safety & Training Specialist/Emergency Response Team Chief
Tucson Electric Power Company,
Springerville Generating Station
St. Johns, Arizona

David Alan Krause, BS, NREMT-P
Safety Management Systems Training Academy
Lafayette, Louisiana

Lawrence Linder, MA, NREMTP
Hillsborough Community College
Tampa, Florida

Matt Marcarelli
Lieutenant, City of New Haven Fire Department
Instructor, Connecticut Fire Academy
Middlesex County Fore School
Northford, Connecticut

Mark Martin
Fire Chief, Perry Township Fire Department
Adjunct Instructor, Stark State College
Stark County, Ohio

Stephen S. Malley
Director of Fire Science Technology
Weatherford College
Weatherford, Texas

Charles W. McDonald
Fire Science Instructor
Trinity Valley Community College
Terrell, Texas

Tracy Rickman
Coordinator, Fire Technology
Rio Hondo College
Santa Fe Springs, California

Thomas Y. Smith, Sr.
Fire Science Program Chair
West Georgia Technical College
LaGrange, Georgia

About the Author

During his twenty-two years in the Los Angeles Fire Department (L.A.F.D.), Gene Mahoney rose through the ranks to become a Battalion Chief of the Watts Battalion, the most active battalion in the department. As a part-time instructor at Los Angeles Junior College, he, and another Los Angeles Fire Department captain, developed a fire science curriculum for the junior college level that is still in use today as the standard in California. At the time of his retirement from the L.A.F.D., he was in command of the Public Relations section of the department.

Prior to his tenure as a Fire Chief, instructor, and program coordinator, Chief Mahoney served in the navy during WWII, was a naval aviation cadet, and spent eighteen years in a naval reserve squadron, retiring as a Lieutenant Commander.

On his off-duty days from the L.A.F.D., Chief Mahoney attended the University of Southern California where he received his B.S. in Public Administration with a minor in Fire Administration. Three years later, Chief Mahoney was awarded a Master's degree in Education.

Mahoney retired from the L.A.F.D in 1969 to accept the position of Fire Chief for the City of Garden Grove. Shortly after, he was promoted to the position of Public Safety Director and placed in command of both the fire and police departments. Eventually his position was upgraded to Assistant City Manager for Public Safety. He retired in 1972 to become the Fire Chief for the City of Arcadia.

Retiring from the active portion of the fire service in 1975, Mahoney then became the Fire Science Coordinator for Rio Hondo College in California. As the program coordinator at Rio Hondo, he expanded the Fire Academy, which then prepared students for qualifications as a Firefighter I in the state of California. He retired from Rio Hondo College in 1988 as a Professor of Fire Science.

Mahoney's writing career began during his time on the L.A.F.D. where he wrote various articles for the *Firemen's Grapevine* and other professional magazines. While assigned to the Training Division of the department, he wrote most of the drill guides used to train apparatus operators and later a special book for the department called *The Progressive Years*. He also wrote one novel, *Anatomy of an Arsonist*, but soon learned that technical books were his calling. Since then, he has published more than twelve fire science textbooks covering topics from hydraulics to fire apparatus and equipment to fire suppression practices and procedures.

In addition to having an accomplished, multifaceted career, Mahoney has also served as a president or chairman for many organizations, including the Boy Scouts of America, United Way, Salvation Army, International Association of Toastmasters, and president of the Rio Hondo College Faculty Association.

Introduction

This book is much different from many of the mathematical books you will find. Most are written in a formalized style, whereas this book is written in a "friendlier," more casual format, and problems are presented in several different ways, hopefully to reach students with varying learning styles.

The material in the book has been prepared especially to meet the needs of members of the fire service. In addition to material that applies directly to the fire service, it also contains information that is useful in situations that fire department members face in their day-to-day life. A number of shortcuts are introduced that prepare individuals to solve various mathematical calculations mentally without the need to refer to a calculator. This is both a useful situation and also extremely satisfying.

CHAPTER 1
INTRODUCTION TO THE WORLD OF NUMBERS

KEY TERMS

abstract	mentally	real numbers
arithmetic	metric system	symbol
composite numbers	negative integers	word problems
customary system	period	zero
digit	positive integers	
integers	prime numbers	

OBJECTIVES AND GOALS

The objectives and goals of this chapter are to introduce the reader to the world of numbers. Upon completion of this chapter, the reader should:

- Understand the basic concepts of numbers.
- Be able to define real numbers, integers, period, prime numbers, and composite numbers.
- Be able to determine whether or not a number is a prime number or a composite number.
- Be able to explain the divisibility rules for 2, 3, and 5.

INTRODUCTION

One of the objectives of this chapter is to introduce the reader to the world of numbers and to define some of the words introduced in this chapter. However, this chapter also has two other very important objectives:

1. To establish a firm foundation for the reader to be able to use the canceling format in solving fraction problems and formulas. To develop this foundation, the reader must have a firm understanding of prime numbers and be able to determine whether or not a number is a prime number.
2. To establish a firm foundation for learning an effective method of reducing a fraction to its lowest terms. This capability will be extremely important in the chapter on fractions. The system to be introduced will require a firm understanding of prime numbers.

WORDS TO REMEMBER

abstract—Not related to a concrete object.

arithmetic—The science or art of computation by positive, real numbers.

composite numbers—A positive whole number greater than one is either a prime number or a composite number. A composite number is a whole number derived from multiplying two or more prime numbers. For example, 2 and 3 are both prime numbers. Multiplying $2 \times 3 = 6$. Therefore, 6 is a composite number composed by multiplying the prime number 2 by the prime number 3.

digit—One of the Arabic numbers, 0 to 9, used to represent numbers in the customary system.

integers—Whole numbers that are positive or negative and do not have any fractional or decimal parts.

mentally—Carried out in the mind without the use of any physical aid.

negative integers—Numbers such as: $-1, -2, -3$, and so on.

period—A group of three digits used for counting purposes.

positive integers—Generally referred to as the counting numbers.

prime numbers—Whole numbers greater than one that can be divided only by themselves or one.

real numbers—Includes whole numbers, integers, fractions, and decimal fractions.

symbol—A written or printed mark representing something abstract.

zero—The integer 0 is neither positive nor negative.

Suggestions for Readers

This book reviews basic math and introduces some new concepts. It is hoped that when the reader finishes the book, he or she will be able to solve math problems without using a calculator or other aids. Consequently, to obtain the maximum benefit from this text, the reader should put away his or her calculator and not refer to it or any other digital instrument for any mathematical calculation. All mathematical problems provided in the text should be solved mentally or with pencil and paper.

Each chapter contains a number of practice problems for the reader to solve. In order to obtain the maximum benefit from this book, it is recommended that the reader solve all of the practice problems. Solutions to the problems are provided in the Reader's Practice Problem section of the Appendix.

The Basic Concept of Numbers

Mathematics may be thought of as the science of operating with, or manipulating, numbers. Mathematics covers a wide scope of subjects that are normally taught as individual courses in high school and college. It includes such subjects as arithmetic, algebra, geometry, trigonometry, analytical geometry, and calculus. Although this book includes a scattering of information that touches on various phases of these assorted mathematics subjects, it primarily concentrates on arithmetic.

Webster defines *arithmetic* as "the science of or art of computation by positive, real numbers, the fundamental operations of which are performed by addition, subtraction, multiplication, and division." The word arithmetic itself comes from two Greek words:

"arithmos," which means number, and "techne," which means art or skill. However, to many adults, including many firefighters, arithmetic is a science or art designed to confuse and frustrate anyone who attempts to comply with the rules for completing its functions. To those who fall into this category, arithmetic is one of the most complicated aspects of life. They had a distaste for it as a child, and the distaste continued into adulthood.

To others, however, it is a stimulating set of rules that challenges them to reduce solutions to their own level. It has positive rules such as $2 + 2 = 4$ and $3 \times 3 = 9$; rules that are carefully structured and lead to positive results. Nothing is more challenging or rewarding to these individuals than finding a method of outwitting a traditional method of solving a problem.

Regardless of which category an individual falls in, numbers are as much a part of life as walking and sleeping. There are some numbers that immediately provide a picture to an individual. Take a look at the following numbers to see what picture each one provides for you.

1776
SAN FRANCISCO, 1906
DECEMBER 7, 1941
9/11

The day a firefighter is to report to duty begins with a number on a clock and an alarm that says "it's time to get up." A few minutes later, she is probably in her car and heading for a duty station. She might stop at a gas station and fill up her vehicle with 87-octane gasoline. She may not know exactly what 87-octane refers to but she uses it because the manufacturer of the vehicle recommends that it be used.

As the firefighter leaves the gas station, her thoughts might turn to the latest report she heard on the radio regarding a Class 3 hurricane due to come ashore in their area later that day. The wind outside is already increasing in strength, and she knows that she will likely be responding to one of the first cries for help. Yes, numbers are a vital part of the life every firefighter.

Introduction to the Number System

A number system refers to counting. Counting in a given part of the world is as old as speech. Strangely, in some areas numbers are older than the written language.

The history of numbers dates back to the beginning of civilization. It appears likely that tallying was among the earliest methods of keeping a record of quantities. For example, a wolf bone was discovered that had a number of notches carved in it that seemed to indicate that an individual had created a tally of some kind. What type of tally is not known. However, the wolf bone dated back to about 30,000 B.C.

Although the individual pebbles had no relationship to a known number, a rancher may have created a pile of pebbles with a single pebble indicating a cow or sheep that had been let out to pasture in the morning. As the evening hour approached and the animals started returning to the pen, a pebble was removed from the pile for each animal that returned. If any pebbles were left in the pile when it appeared that all animals had returned, the farmer would know that there were still animals out in the field.

As civilization emerged from the farm to building cities, the number of commercial transactions between individuals and countries increased. The increase in commercial transactions demanded that a system for keeping records be developed. This means that a number system becomes essential. Records show that the ancient Egyptians developed a number system to keep records of what was bought and sold. However, the Egyptian civilizations used their knowledge of numbers for more than keeping records of what was bought and sold. There should be little doubt in anybody's mind who is familiar with construction that the individuals who developed and built the pyramids more than 5,000 years ago had an amazing grasp of numbers. The blocks used to build the pyramids were cut with very little space between them. The walls inside the pyramids were filled with hieroglyphic symbols indicating that not only did the builders of the pyramids understand their number system, but so did those who read the artwork.

The Egyptian number system is a hieroglyphic system written from right to left. A hieroglyphic system uses pictures to represent words. The system was based on the decimal system. This means that it had separate symbols for the one unit, ten unit, hundred unit, thousand unit, ten thousand unit, hundred thousand unit, and one million unit. However, it contained no zero or principle of position.

In addition to the Egyptian number system, many systems were developed in various parts of the world. Some of the systems that have been identified are:

- Egyptian
- Babylonian
- Chinese
- Roman
- Greek
- Hebrew
- Arabic
- Hindu
- Sumerian
- Mayan
- Hindu–Arabic

However, despite the development of so many numbering systems, only two primary numbering systems are used throughout the world today: the ***metric system*** and the ***customary system***. The basis for the customary system dates back to the ancient Hindu–Arabic system. The customary system is officially used only in the United States and two other countries, Liberia and Myanmar (formerly Burma). It is the system that will be used for most of the problems in this book. However, the metric system is referred to in several chapters.

An Explanation of Numbers

real numbers—Includes whole numbers, integers, fractions, and decimal fractions.

integers—Whole numbers that do not have any fractional or decimal parts. Integers may be positive or negative and includes zero.

positive integers—Generally referred to as the counting numbers. Examples are 1, 2, 3, 4, and so on.

negative integers—Numbers such as $-1, -2, -3$, and so on. All negative integers are less than zero.

zero—The integer 0 is neither positive nor negative.

period—A group of three digits used for counting purposes. For example, the number 437,794,423 is composed of three periods. The three periods are referred to as the millions period (437), the thousands period (794), and the hundreds period (423).

prime numbers—Numbers greater than one that can be divided evenly by only themselves or one.

composite numbers—A whole number is either a prime number or a composite number. A composite number is a whole number derived from multiplying together two or more prime numbers. For example, 2 and 3 are both prime numbers. However, $2 \times 3 = 6$, which makes the 6 a composite number.

Determining Prime Numbers

Prime numbers, also known as primary numbers, are an important part of the customary system. They are used effectively to determine when a fraction has been reduced to its lowest term and are also important when working with formulas. Primary numbers are the basis element, taking advantage of canceling fractions or portions of formulas. Knowledge of, and being able to use, primary numbers should be considered a primary step in working with numbers.

The following method can be used to determine all prime numbers from 2 to 100.

Write out all the numbers from 2 to 100 on a separate piece of paper in the following fashion:

```
              2,  3,  4,  5,  6,  7,  8,  9, 10
11, 12, 13, 14, 15, 16, 17, 18, 19, 20
21, 22, 23, 24, 25, 26, 27, 28, 29, 30
31, 32, 33, 34, 35, 36, 37, 38, 39, 40
41, 42, 43, 44, 45, 46, 47, 48, 49, 50
51, 52, 53, 54, 55, 56, 57, 58, 59, 60
61, 62, 63, 64, 65, 66, 67, 68, 69, 70
71, 72, 73, 74, 75, 76, 77, 78, 79, 80
81, 82, 83, 84, 85, 86, 87, 88, 89, 90
91, 92, 93, 94, 95, 96, 97, 98, 99, 100
```

The first three prime numbers are 2, 3, and 5. Therefore, any of the other 100 numbers that can be divided evenly by any one of these three numbers is not a prime number.

Divisibility Rules

There are three divisibility rules that are useful for determining whether or not a number is a prime number:

- A number is divisible by 2 if it is an even number. Examples of even numbers are 8, 22, 56, 78.
- A number is divisible by 3 if the sum of its digits is divisible by 3. Examples of numbers divisible by three are 9, 27, 54, 81.
- A number is divisible by 5 if it ends in 0 or 5. Examples of numbers divisible by 5 are 10, 25, 50, 75.

Using these three divisibility rules, cross out any of the numbers on the chart you developed that can be divided by 2, 3, or 5. Check your results against the following chart.

Those numbers that are highlighted in light gray are divisible by 2 or 5; those highlighted in dark gray are divisible by 3.

```
              2,  3,  4,  5,  6,  7,  8,  9, 10
             11, 12, 13, 14, 15, 16, 17, 18, 19, 20
             21, 22, 23, 24, 25, 26, 27, 28, 29, 30
             31, 32, 33, 34, 35, 36, 37, 38, 39, 40
             41, 42, 43, 44, 45, 46, 47, 48, 49, 50
             51, 52, 53, 54, 55, 56, 57, 58, 59, 60
             61, 62, 63, 64, 65, 66, 67, 68, 69, 70
             71, 72, 73, 74, 75, 76, 77, 78, 79, 80
             81, 82, 83, 84, 85, 86, 87, 88, 89, 90
             91, 92, 93, 94, 95, 96, 97, 98, 99, 100
```

If you completed the operation correctly, leaving the 2, 3, and 5, the following numbers should remain on your chart.

```
        2, 3, 5, 7, 11, 13, 17, 19, 23, 29, 31, 37,
        41, 43, 47, 49, 53, 59, 61, 67, 71, 73, 77,
        79, 83, 89, 91, 97
```

The next number of those remaining above 5 that is not divisible by 2, 3, or 5 is 7. Therefore, it is a prime number. The remaining numbers above 7 that are evenly divisible by 7 are 49, 77, and 91. These should be removed from the list.

```
        2, 3, 5, 7, 11, 13, 17, 19, 23, 29, 31, 37,
        41, 43, 47, 49, 53, 59, 61, 67, 71, 73, 77,
        79, 83, 89, 91, 97
```

The remaining numbers after the 49, 77, and 91 have been removed are prime numbers. It can therefore be concluded that the prime numbers from 2 to 100 are:

```
        2, 3, 5, 7, 11, 13, 17, 19, 23, 29, 31, 37, 41,
        43, 47, 53, 59, 61, 67, 71, 73, 79, 83, 89, 97
```

Determining Whether a Number Larger than 100 is a Prime Number

Occasionally, it is desired to determine whether a number larger than 100 is a prime number. The method to determine this is relatively easy to use.

Problem: Is 197 a prime number?

Solution: The first step is to determine the largest number that when squared (multiplied by itself) is smaller than the number being examined (197). It happens to be 14 (14 × 14 = 196).

The next step is to determine whether any of the prime numbers smaller than 14 can be divided evenly into 197. The prime numbers smaller than 14 are: 2, 3, 5, 7, 11, 13.

- 2 cannot be divided evenly into 197 because 197 is not an even number.
- 3 cannot be divided evenly into 197 because the sum of its digits is 17, which cannot be divided evenly by 3.
- 5 cannot be divided evenly into 197 because 197 does not end in a 5 or zero.
- 7 cannot be divided evenly into 197.
- 11 cannot be divided evenly into 197.
- 13 cannot be divided evenly into 197.

Conclusion: Because none of the prime numbers smaller than 14 can be divided evenly into 197, 197 is a prime number.

Problem: Prove that 271 is a prime number.

Solution: The first step is to determine the largest number that when squared (multiplied by itself) is smaller than the number being examined (271). 16 × 16 = 256 whereas 17 × 17 = 289. Therefore, 16 is the largest number, which, when squared, is smaller than 271.

The next step is to determine whether any of the prime numbers smaller than 16 can be divided evenly into 271. The prime numbers smaller than 16 are: 2, 3, 5, 7, 11, 13.

- 2 cannot be divided evenly into 271 because 271 is not an even number.
- 3 cannot be divided evenly into 271 because the sum of its digits is 10, which cannot be divided evenly by 3.
- 5 cannot be divided evenly into 271 because 271 does not end in a 5 or zero.
- 7 cannot be divided evenly into 271.
- 11 cannot be divided evenly into 271.
- 13 cannot be divided evenly into 271.

Conclusion: Because none of the prime numbers smaller than 16 can be divided evenly into 271, 271 is a prime number.

It is important to remember this process because it may be valuable when reducing fractions to their lowest term and when canceling.

The value of understanding prime numbers will be demonstrated later in the book when working with fractions and in sections where canceling numbers can be used to an advantage.

Reader's Practice Problems—1.1

1. What is the largest number that when squared is less than 173?
2. What prime numbers should be used to determine whether 173 is a prime number?
3. Is 173 a prime number?
4. Prove that 113 is a prime number.
5. Prove that 199 is a prime number.

The answers are provided in the Reader's Practice Problems section of the Appendix.

Positive and Negative Integers

Some people have difficulty understanding the concept of positive and negative integers, especially when problems are encountered requiring that they be added, multiplied, subtracted, or divided. Probably the best way to understand them visually is by the use of a line chart.

Integers can be thought of as lines on the chart. The zero on the chart is neither a negative nor a positive integer.

$$-5\ -4\ -3\ -2\ -1\ \ 0\ \ +1\ +2\ +3\ +4\ +5$$
Negative integers   Positive integers

The concept can be cleared up better for some individuals by using a real-life situation. Imagine that your net worth depends entirely on two factors—your bank account and your credit card. Your bank account can be considered as a positive integer, whereas the amount of money you owe on a credit card account can be considered as a negative integer.

If you subtract the amount of one of the two from the other you will determine your net worth. If the amount in your bank account is larger, you will have a positive net worth. If the amount you owe on your credit card account is larger, you will have a negative net worth. Problems involving addition will be examined in Chapter 2.

Word Problems

Problems are provided at the end of each chapter for the reader to solve. Some of the problems are in typical mathematical formats such as:

$$\begin{array}{cccc} 345 & 892 & 43 & 643 \div 17 \\ 189 & -453 & \times 37 & \\ +764 & & & \end{array}$$

Others are presented as **word problems**. With word problems, it is up to the reader to arrange the problem in the proper form for solving. Regardless, for maximum benefit to the reader, all problems is this book should be solved mentally or with pen or pencil.

Some word problems require a single operation to solve, such as adding, multiplying, or dividing. Others contain material that will require two or more operations.

Word problems can be tricky. Following are some suggestions for solving them.

1. First, completely read the problem.
2. Determine what the problem is asking to be solved.
3. Reread the problem and select those factors that are needed to solve the problem.
4. When possible, develop a numerical formula that will provide the answer that is required.

The first step is a very important one. For example, try to solve the following problem as quickly as you can. Imagine that it is given to you verbally.

Problem: A bus driver left the station with five passengers onboard. At the first stop, three people got on and two got off. At the next stop, four got on and two got off. At the following stop, five got off and two got on. Three got off at the next stop and two got on. The same amount got on and off at the following stop.

Question: How many stops did the bus driver make?

If this problem had been given to you verbally, you probably would have started adding and subtracting numbers at the first stop. Most people do. However, this problem should illustrate the importance of reading the entire problem prior to trying to solve any portion of it.

The second item regarding word problems is to recognize that items may be included in the wording that have nothing to do with solving the problem. Take a look at the following example. First, determine what the problem is asking to be solved. Then select the items from the problem that are needed to solve the problem.

Problem: The cook at Fire Station Two bought fruit every time he went shopping for the evening meal. On his first trip to the market he bought five apples, four oranges, and two peaches. On his next trip, he bought three apples, six oranges, and four peaches. His purchase on the third visit was three apples, two oranges, and four peaches. On his fourth visit, he bought six apples, six oranges, and six peaches. What was the total amount of oranges and apples he bought during his four visits to the store?

Solution: The question to be solved is to determine the total number of oranges and apples purchased in his four visits to the store. During the first visit, the number of apples and oranges purchased was 9. On the second visit, it was also 9. On the third visit he purchased 5, and on the fourth visit he bought 12. The word formula would be:

Total apples and oranges purchased on his four visits = number purchased on first visit + number purchased on second visit + number purchased on third visit + number purchased on fourth visit.

With the information you gathered on rereading the problem, it is possible to set up the following number formula:

$$X = 9 + 9 + 5 + 12 = 35$$

Changing Word Problems to Number Problems

One of the most difficult things for many people is changing the word problem to a number problem. Once this is done, solving the math problem is generally easy.

It is important when changing a word problem to a number problem to work in an organized manner. Working in an organized manner refers to changing the wording in the problem into a numerical equation. One thing to assist in organizing the problem is to recognize that there are certain key words that will provide a hint as to whether to add, subtract, multiply, or divide certain parts of the written word. It is good practice to let X always stand for the answer to the problem.

Sample Word Problems

Most of the word problems that will be encountered in the fire service are provided in a format for which a formula is available. The reader must be familiar with all of the formulas and know which one to use to solve the problem. Following are three examples of how this is normally given.

Problem #1: A circle has a diameter of 10 feet. What is the area of the circle?

Solution: To solve this problem, the reader must know that the formula for determining the area of a circle when the diameter is known is:

$$\text{area} = .7854 D^2$$

The problem states that the diameter of the circle is 10 feet. The 10 feet should be placed into the formula.

$$\begin{aligned} \text{area} &= .7854 D^2 \\ &= (.7854)(10)(10) \\ &= 78.54 \text{ sq ft.} \end{aligned}$$

Note: The area of a circle can also be solved by the formula:

$$\text{area} = \Pi(r^2)$$

However, because most of the fire department formulas for finding the area of a circle are used for finding the area of a nozzle, the diameter rather than the radius is used.

Problem #2: Water is being discharged from a $1\frac{1}{4}$-inch nozzle tip at a nozzle pressure of 50 psi (pounds per square inch). What is the nozzle reaction?

Solution: To solve this problem, the reader must know that the formula for determining the nozzle reaction is:

$$\text{NR (nozzle reaction—in lbs)} = 1.5 \, D^2 P$$

The problem states that the nozzle pressure is 50 psi and the nozzle diameter is $1\frac{1}{4}$-inch. These figures must be placed into the formula.

$$NR = (1.5)(1\tfrac{1}{4})(1\tfrac{1}{4})(50)$$
$$= (1.5)(1.25)(1.25)(50)$$
$$= 117.19 \text{ pounds}$$

Problem #3: What is the volume of a rectangular container that measures 10 feet by 6 feet by 8 feet?

Solution: To solve this problem, the reader must be aware that the formula for finding the volume of a rectangle is:

$$\text{volume} = lwh$$

where l = length
        w = width
        h = height

The problem states that the length is 10 feet, the width is 6 feet, and the height is 8 feet. These figures must be placed into the formula.

$$\text{volume} = (10)(6)(8)$$
$$= 480 \text{ cubic feet}$$

Although hydraulic formulas are provided for determining the answer to most of the word problems that will be presented to firefighters in organized classes or on the fire ground, the format that will be used for presenting the material needed for solving the problem will vary. The first three problems just listed the material in standard word problem format.

However, many of the word problems that will be encountered by firefighters will present the material needed to solve the problem in a photo or drawing. This problem will use this procedure to familiarize the reader with the other format that will be encountered.

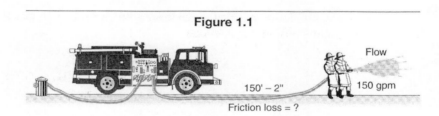

Figure 1.1

Problem #4: What is the friction loss in the Figure 1.1 layout?

Solution: The formula for determining the friction loss is:

$$FL = CQ^2L$$

where $C = 8$
        $Q = 150/100 = 1.5$ (moving the decimal point)
        $L = 150/100 = 1.5$

To solve, place the information in the formula.

$$FL = (8)(1.5)(1.5)(1.5)$$
$$= 27 \text{ psi}$$

Reader's Practice Problems—1.2

1. A 2-inch tip is being used on a portable monitor. What is the area of this tip?

    The formula for determining the area of a circle is

    $$\text{area} = .7854D^2$$

    where   D = the diameter of the tip

2. Water is being discharged from a 1-inch tip that has a nozzle pressure of 80 psi. What is the nozzle reaction on this tip?

    The formula for determining the nozzle reaction on a tip is

    $$\text{nozzle reaction} = 1.5\, D^2 P$$

    where   D = diameter of the tip

    P = nozzle pressure

3. A rectangular container measures 15 feet by 10 feet by 8 feet. What is the volume of this container?

    The formula for determining volume is

    $$\text{volume} = lwh$$

    where   l = length

    w = width

    h = length

The answers are provided in the Reader's Practice Problems section of the Appendix.

REVIEW QUESTIONS

1. What is a digit?
2. What are the two primary numbering systems used throughout the world today?
3. What is the meaning of "real numbers"?
4. What are integers?
5. Is zero a positive or negative integer?
6. What is the definition of a prime number?
7. What is the definition of a symbol?
8. What is a composite number?
9. What are the divisibility rules for 2, 3, and 5?
10. What method is used to determine whether or not a number is a prime number?
11. What is the definition of abstract?

Test 1

WORD PROBLEMS

1. A $1\frac{1}{2}$-inch tip is being used on the end of a portable monitor. What is the area of this tip?
2. Water is being discharged from a 1-inch tip at 75 psi. What would be the nozzle reaction on this tip?
3. A rectangular container measures 15 feet by 12 feet by 10 feet. What is the volume of this container?
4. What is the volume of a rectangular container that measures 9 inches by 6 inches by 8 inches?
5. What is the area of a $2\frac{1}{2}$-inch tip that is being used on the end of a portable monitor?
6. What is the friction loss in a 350-feet hose that has a friction factor of 8 and is discharging 100 gpm of water?

CONVENTIONAL PROBLEMS

For each number below, list which are the prime numbers within that number.

1. 7,239,461
2. 4,622,174
3. 5,386,034
4. 8,843,902
5. 6,783,210
6. 78,945
7. 65,927
8. 49,869
9. 3,479,224
10. 6,245,442
11. 8,763,225
12. 2,567,864
13. 4,562,987
14. 39,748
15. 68,934

Chapter Summary

This chapter was designed to establish the foundation for work in the ensuing chapters. Many words were introduced so readers would understand their meaning as they are encountered later in the book. Readers were also introduced to the concept of numbers and, specifically, prime numbers. Moreover, information regarding working with word problems was also provided, along with some examples The chapter also included review questions and a test; if the reader thoroughly understands all the material presented in the chapter, he or she is prepared to move on to the next chapter.

CHAPTER 2
SOLVING AND PROVING WHOLE NUMBER ADDITION PROBLEMS

KEY TERMS

addend
adding the digits method
casting out nines method
excess of nines
sum

OBJECTIVES AND GOALS

The objectives and goals of this chapter are for the reader to review the addition process, the process of adding positive and negative integers, and to learn how to prove that the answer obtained in an addition problem is correct. Upon completion of this chapter, the reader should:
- Be capable of solving addition problems.
- Be capable of solving addition problems involving positive and negative integers.
- Be able to prove that the answer to an addition problem is correct.

INTRODUCTION

In reality, this chapter, like many of those that follow, is a review chapter. Almost all readers have been exposed at one time or another to the process of adding whole numbers. Unfortunately, however, all readers will not have the same level of achievement in the addition process. Some readers will sincerely appreciate the opportunity for reviewing the process, whereas others will consider it a waste of time. However, if you fall into the second category, it is suggested that you take a good look at the process for proving that an answer obtained in an addition problem is correct. This procedure will most likely be new to most readers. It is a valuable tool that will be used in many addition problems in this book.

WORDS TO REMEMBER

addend—A number or quantity to be added to another number or quantity.

sum—The result obtained by adding together two or more digits or numbers.

$$
\begin{array}{rcl}
347 & \leftarrow & \text{addend} \\
850 & \leftarrow & \text{addend} \\
\underline{563} & \leftarrow & \text{addend} \\
1{,}760 & \leftarrow & \text{sum}
\end{array}
$$

A Review of the Addition Process

Addition problems may be presented as: Add 2 and 3; 2 + 3; or $\begin{array}{r}2\\+\ 3\end{array}$

Adding two digits and obtaining an answer is referred to as the addition process. The addition process is generally regarded as a positive science. For example, 2 plus 2 is always 4, and 8 plus 6 is always 14. However, sometimes a problem does not turn out to be what it appears to be. Be very careful when working word problems that you do not fall into this trap. Following is an example of what appears to be a word problem but in reality is a mystery word problem. It is not related to a fire service or emergency medical care situation but is a good example of a word problem that is not what it first appears to be.

A man went into a hotel and asked about the price of a room. The clerk replied that the rate was $10 per night. "I would like a room for three nights," the man said. The man paid the required thirty dollars. The clerk logged him in and gave him the keys to the room. After the man went up to his room, the clerk discovered that the rate for the three nights was only $25. He gave the bellhop five one-dollar bills and told him to take them up to the man's room.

On the way up to the room, the bellhop decided he would give the man only three of the bills. So he stuck two in his pocket. Now the problem is:

The man paid $27 for the room. The bellhop has $2. What happened to the other dollar?

If you can remember that far back, it will dawn on you that you learned how to add in elementary school. While learning, you probably used your fingers, and you may have also used some matchsticks provided by the teacher. However, learning to add was much more different than having to memorize a series of tables like you did when learning to multiply. In fact, there probably weren't any addition tables at that time. If there had been one, it most likely would have looked like the following:

	1	2	3	4	5	6	7	8	9	10
1	2	3	4	5	6	7	8	9	10	11
2	3	4	5	6	7	8	9	10	11	12
3	4	5	6	7	8	9	10	11	12	13
4	5	6	7	8	9	10	11	12	13	14
5	6	7	8	9	10	11	12	13	14	15
6	7	8	9	10	11	12	13	14	15	16
7	8	9	10	11	12	13	14	15	16	17
8	9	10	11	12	13	14	15	16	17	18
9	10	11	12	13	14	15	16	17	18	19
10	11	12	13	14	15	16	17	18	19	20

Select a digit in the left-hand vertical column that you wish to add to a digit in the top horizontal column, or select a digit in the top horizontal column that you wish to add to a digit in the left-hand vertical column. Where the two intersect provides the answer to your addition problem. For example, you wish to add 7 + 6, select the 7 from the left-hand vertical column and the 6 from the top horizontal column, or the 7 from the horizontal column and the 6 from the left-hand vertical column. Note that they intersect at 13. Therefore, 7 + 6 = 13.

The addition chart is quite simple, isn't it? In fact, you will probably have very little problem constructing it from memory. You will be tested later in this chapter to see if this is true.

Now take a look at the following addition problem.

$$\begin{array}{r} 6 \\ 9 \\ 5 \\ 3 \\ \underline{2} \end{array}$$

It is doubtful that you would have any trouble solving it.
However, take a look at this one.

$$\begin{array}{r} 83,462 \\ 49,936 \\ 80,423 \\ 67,377 \\ \underline{60,886} \end{array}$$

You probably think this problem is more difficult than the previous one. However, isn't it just five of the first problem when divided up as illustrated?

$$\begin{array}{ccccc} 8 & 3 & 4 & 6 & 2 \\ 4 & 9 & 9 & 3 & 6 \\ 8 & 0 & 4 & 2 & 3 \\ 6 & 7 & 3 & 7 & 7 \\ \underline{6} & \underline{0} & \underline{8} & \underline{8} & \underline{6} \end{array}$$

That's exactly what an addition problem is: a series of single columns added individually. Each column in a problem has been given a name based on its location in the problem. The names of the columns are identified in the following problem.

$$\begin{array}{ccccc} 4 & 7 & 3 & 2 & 6 \\ 5 & 2 & 2 & 6 & 1 \\ \underline{3} & \underline{0} & \underline{7} & \underline{5} & \underline{1} \end{array}$$

ten thousands	thousands	hundreds	tens	ones

Now work the problem.

$$\begin{array}{ccccc} 1 & 1 & 1 & & \\ 4 & 7 & 3 & 2 & 6 \\ 5 & 2 & 2 & 6 & 1 \\ \underline{3} & \underline{0} & \underline{7} & \underline{5} & \underline{1} \\ 13 & 0 & 3 & 3 & 8 \end{array}$$

The ones column is added first. It totals 8. Because the sum is 9 or less, it is placed below the ones column. If it is greater than 9, such as it will be in the tens column, then the right-hand digit is placed in the answer in the tens column and the left-hand digit is carried over and added to the hundreds column. The tens column totals 13. The 3 is placed in the answer in the tens column and the 1 is carried over and added to the digits in the hundreds column. With the 1 that is carried over, the total of the hundreds column is 13. The 3 is added to the hundreds column and the 1 is carried over to the thousands column. The thousands column now totals 10. The 0 is added to the thousands column and the 1 is carried over to the ten thousands column. This column now totals 13. The 13 is added to the answer. The answer to the problem is now 130,338.

Let's take you back a few years and see if you still remember what you learned in elementary school. Work each of the following problems and place your answer in the space provided. All of the problems are addition problems.

One-digit problems:

```
4   7   2   4   8   6
5   4   9   3   7   2
```

Two-digit problems:

```
56   35   87   48   39   66   93   74   69   70
35   76   37   94   38   93   23   47    6   34
                         32   86   45   34   84
```

Three-digit problems:

```
649   569   271   326   473   675   284   319   759
251   987   528   537   556   362   835   832   636
                              341   753   664   342
```

Arrange the following into vertical columns and add:

$$5,924 + 5,832 + 3,721$$
$$8,793 + 3,462 + 7,329 + 3,461$$
$$4,381 + 7,329 + 5,438 + 5,382 + 3,482$$

It is now time to see how well you did on the above problems. Take the following Reader's Practice Problems and check your results with the answers found in the Appendix.

Reader's Practice Problems—2.1

Solve the following addition problems.

One-digit addition problems:

```
3   5   4   7   9
4   9   4   8   9
```

Two-digit addition problems:

```
23   78   54   32   97   13   65   92   46
72   67   98   27   89   75   26   18   43
                         66   14   94   75
```

Three-digit addition problems:

```
347   937   237   556   783   564   238   563   544
666   316   983   723   437   391   733   239   663
                              743   968   453   867
```

Arrange the following into a vertical problem and add.

$$4,327 + 9,563 + 6,632$$
$$3,605 + 9,745 + 7,702 + 4,863$$
$$8,854 + 6,729 + 5,623 + 3,398 + 4,582$$

You probably noticed that the answers to the problems you worked on prior to taking the Reader's Practice Problem test were not provided. However, there is always a question that remains after any mathematic problem has been solved: Is the answer correct? There is little value in working a problem without having some method of determining whether or not the answer is correct. No doubt you were provided some method of checking problems as you wandered through the mathematical learning curve. However, some of the methods were strung out and sometimes difficult to use.

The method used in this book is based upon a principle referred to as the "excess of nines" method. This method is probably completely new to most of you; however, for those who master it there will be times that it will prove extremely beneficial.

Proving Addition Problems

Now let's assume that you are taking an examination for promotion, or perhaps to obtain a new position. Let's also assume that the instructions for the examination stated that in order to continue in the examination process the answer you obtain on this problem must be correct. In other words, if your answer to the problem is wrong, you will not receive the promotion or the job. Your answer to the problem is the same as the one shown below. What would you do now?

$$\begin{array}{r} 62,234 \\ 35,721 \\ 43,893 \\ 56,387 \\ \underline{21,446} \\ 219,681 \end{array}$$

If you are like most people, and it is assumed you are, you would probably work the problem again. If you obtained the same answer the second time, you might accept it as being correct. However, because of the importance of getting it right, you might even decide to work it a third time.

However, if you obtained a different answer the second time than you did the first time, there is little doubt that you would fail to work it a third time. If the answer on the third try was the same as one of the first two answers, you might accept the answer that matches your third operation or even work it again. Regardless, whatever you do will take up a lot of time, and if the time allotted for completing the examination is limited, time will be valuable.

The procedure to use for checking an addition problem to see whether it has been solved correctly is based on the "excess of nines" principle, which is much easier to use than working the problem over. Following is an illustration of how the procedure works.

$$\begin{array}{r} 62,234 \\ 35,721 \\ 43,894 \\ 56,587 \\ \underline{21,446} \\ 219,882 \end{array}$$

One of the primary objectives of this chapter is to introduce the tool needed to use a simple method of checking any addition problem. The procedure can also be used to prove that a subtraction, multiplication, division, or square root problem has been worked correctly. However, in this chapter it will be used only for checking addition problems. This is extremely important because it is necessary for an individual not only to solve a mathematical problem but to also solve it correctly. At times, an incorrect answer can be dangerous. A good example is working a problem to determine the pressure that should be supplied on a pumper to achieve the desired results at the nozzle. Any firefighter who has been on the end of a working hose line that is supplied by too much pressure is well aware of the result.

After working the example problem, the first step to take to determine whether the answer is correct is to determine the "excess of nines" of each of the addends that were added.

The *excess of nines* is a number ranging from one up to and including eight that remains in a two-digit or larger number after all nines, or combination of numbers totaling nine, are removed from the number. The principle is also referred to as the *"casting out nines"* method. For example, in the number 390, if the 9 is removed, the numbers remaining would be 3 and 0: $3 + 0 = 3$. Therefore, the excess of nines in the number 390 is 3.

Another example:
In the number 4,561, the $4 + 5 = 9$. If the 4 and 5 are removed from the number, the remaining numbers are 6 and 1: $6 + 1 = 7$. Therefore, the excess of nines in the number 4,561 is 7.

Some larger numbers have more nines, or combinations of digits equal to nine, in them. In this case, all nines must be removed to determine the excess of nines. For example, examine the number 206,915. Note that there is a 9 in the number and that the $2 + 6 + 1 = 9$. The numbers remaining when 2, 6, 9, and 1 are removed are the 0 and the 5: $0 + 5 = 5$. Therefore, the excess of nines in the number 206,915 is 5.

Other large numbers do not contain either a 9 or a combination of digits that directly total 9. In this case, it is necessary to borrow a digit from one of the other digits in the number to add to other digits in order to obtain a 9. For example, examine the number 2,248. Note that there is not a 9 in the number and no combination of digits in the number directly total 9. However, it is possible to remove a 1 from the 4 (leaving 3) and add it to the 8 to obtain a 9. If you removed a 1 from the 4, then the remaining digits after the 9 is removed would be 2, 2, and 3: $2 + 2 + 3 = 7$. Therefore, the excess of nines of the number 2,248 is 7.

At times, none of the digits in the addition problem equal 9. When this happens, all of the digits in the number can be added. This method is referred to as the *"adding the digits"* method. For example, look at the number 34,177. Note that none of the digits can be added to obtain 9. In this case, add the digits. The result would appear as: $3 + 4 + 1 + 7 + 7 = 22$. Then add the $2 + 2 = 4$. Four would be the "excess of nines" of the number.

Now, let's figure out the number of "excess of nines" that were in each of the digits that were added in the problem we were solving. The problem was:

$$\begin{array}{r} 62,234 \\ 35,721 \\ 43,894 \\ 56,587 \\ \underline{21,446} \\ 219,882 \end{array}$$

The digits that were added in the problem were:

$$62,234$$
$$35,721$$
$$43,894$$
$$56,587$$
$$\underline{21,446}$$

Let's determine the number of "excess of nines" in each of the digits that were added.

First, the 62,234

$6 + 2 + 2 + 3 + 4$ = Take out the 2, 3, & 4
$6 + 2 = 8$ This leaves the $6 + 2 = 8$
The 8 is the "excess of nines."

Second, the 35,721

$3 + 5 + 7 + 2 + 1$ = Take out the 3, 5, & 1
= the 7 & 2
All of the numbers have been removed. The "excess of nines" is zero.

Third, the 43,894

Take out the 9
$4 + 3 + 8 + 9 + 4 = 28$
This leaves $4 + 3 + 8 + 4 = 19$
$19 = 10 + 9$
Take out the 9
This leaves 10
Take a 9 from the 10
This leaves 1. 1 is the "excess of nines."

Fourth, the 56,587

$5 + 6 + 5 + 8 + 7$ None of the digits add to 9. Therefore, add all the digits.
$5 + 6 + 5 + 8 + 7 = 31$
$3 + 1 = 4$

This leaves 4. 4 is the "excess of nines."

Fifth, the 21,446

$2 + 1 + 4 + 4 + 6$ = Take out the 1, 4, & 4
$2 + 6 = 8$ This leaves $2 + 6$
$2 + 6 = 8$. 8 is the "excess of nines."

The next step is to determine the "excess of nines" of the addends.
They were 8, 0, 1, 4, and 8.
Add them to determine the "excess of nines" of the problem.

$$8 + 0 + 1 + 4 + 8 = 21$$
$$2 + 1 = 3$$

Therefore, the "excess of nines" of the addends is 3. ←←←←←←←←←←←
                                                                    ↑
                                                                    ↑
                                                                    ↑
The next step is to determine the "excess of nines" of the answer. ← same
                                                                    ↓
                                                                    ↓
The answer for the problem was 219,882   ↓
                                                                    ↓
                                                                    ↓
Add the digits. $2 + 1 + 9 + 8 + 8 + 2 = 30$ ↓
                   $3 + 0 = 3$ ←←←←←←←←←←←←←←←←←←

If the "excess of nines" of the addends and the answer are the same, the problem is considered to have been worked correctly. Although there is a slight possibility that the two might be the same and the answer wrong, it is so slight that it is generally ignored. In order for it to be the same, errors totaling exactly nine would have had to be made. However, one thing is certain: If the two are different, the answer is definitely wrong. With the illustrated problem, they were both the same.

Following is a suggestion of how to record the information obtained from the three steps taken to prove that this addition problem has been worked correctly.

```
 62,234   — 8   First, add the "excess of nines" of the
 35,721   — 0   numbers in the problem.
 43,894   — 1   8 + 0 + 1 + 4 + 8 = 21
 56,587   — 4                       2 + 1 = 3 ← same ← ↓
 21,446   — 8                                         ↓
219,882   — 2 + 1 + 9 + 8 + 8 + 2 = 30    3 + 0 = 3
          Second, add the numbers in the answer.
```

Clues to Errors

1. If the "excess of nines" of the answer is 1 larger than the "excess of nines" of the problem, one of the digits in the answer is probably one digit too large.
2. If it is 2 greater, then one of the digits in the answer is probably 2 too large or perhaps two of the digits are each one larger than they should be.
3. If the difference is 3 or more, then the error could be in a single digit, or in two or more digits with the total of the errors equaling the difference in the "excess of nines."
4. If the "excess of nines" of the answer is 1 less than the "excess of nines" of the problem, one of the digits of the answer is probably one digit too small.
5. If it is 2 less, then one of the digits in the answer is probably 2 too small or perhaps two of the digits are each one less than they should be.
6. If the difference is 3 or more, then the error could be in a single digit, or in two or more digits with the total of the errors equaling the difference in the "excess of nines."

The analysis used in the rules for errors in addition problems can also be applied for errors for proving answers in other mathematical calculations. However, in this book the traditional method of determining whether or not the problem has been worked correctly will be used for most other mathematical calculations.

Proving Smaller Addition Problems

The previous method for proving an addition problem was based on a multistep process. First, the "excess of nines" of each of the sets of digits being added was determined. Then the "excess of nines" of the answer to the problem was determined. For smaller addition problems, these steps can be combined into a single step when desired, as shown in the following examples.

Problem: Determine the "excess of nines" in the following addition problem.

$$\begin{array}{r} 567 \\ 423 \\ \underline{178} \\ 1,168 \end{array}$$

Solution: Think of the problem as one set of digits.

$$567 \quad 423 \quad 178$$

Eliminate the 6 & 3, the 7 & 2, the 4 & 5, and the 1 & 8.

$$\begin{array}{ccc} \downarrow\downarrow\downarrow & \downarrow\downarrow\downarrow & \downarrow\downarrow \\ 567 & 423 & 178 \end{array} \quad \text{------ eliminate}$$

Answer: 7 (it is the only number left).

The answer to the problem is 1,168. The "excess of nines" of 1,168 is also 7. Therefore, it can be concluded that the problem has been worked correctly.

Problem: Determine the "excess of nines" in the following addition problem.

$$\begin{array}{r} 7,146 \\ 2,584 \\ \underline{1,698} \\ 11,428 \end{array}$$

Solution: Think of the problem as one set of digits.

$$7146 \quad 2584 \quad 1698$$

Eliminate the 7 & 2, the 4 & 5, the 9, the 1 & 8, and the 1 & 8.

$$\begin{array}{ccc} \downarrow\downarrow\downarrow & \downarrow\downarrow\downarrow & \downarrow\downarrow\downarrow \\ 7146 & 2584 & 6198 \end{array} \quad \text{------ eliminate}$$

The remaining numbers are: 6 + 4 + 6 = 16
1 + 6 = 7 ←←←←←←
                ↓
              same
The "excess of nines" of the answer (11,428) is also 7.←← ↑

How large an addition problem can this system be used on? There is no set answer. It is entirely up to the problem solver. Try it. You'll like it. You will particularly like it on some of the practical hydraulic problems.

Reader's Practice Problems—2.2

Check each of the following problems by the "excess of nines" method to determine whether it is correct or wrong. If it is correct, move to the next problem and check it. If it is wrong, work it and when your answer is correct, move on to the next problem.

34	97	56	326	524	785	671
23	44	87	567	742	136	287
57	141	142	893	1,256	555	568
					1,576	1,245

5,372	6,738	1,249	6,732	6,288
6,734	8,495	6,666	5,378	7,632
5,537	6,743	8,967	4,386	3,481
4,387	3,156	3,298	8,743	8,223
11,130	24,132	20,180	25,239	24,524

The answers are in the Reader's Practice Problems section of the Appendix.

Introduction to Positive and Negative Integers

The information in the first part of this chapter for adding whole numbers was limited to adding positive whole numbers because the fire service generally operates within the customary numbering system. However, the Emergency Medical Service not only operates within the customary numbering system but also devotes some time to the metric system. The metric system uses a large number of negative numbers. Consequently, it is advisable that all readers of this book be familiar with both the positive and negative number system.

First, it is essential that all readers understand that the word "integers" refers to positive and negative whole numbers that have no fraction or decimal parts. *Positive whole numbers* refer to the counting numbers such as 1, 2, and 3. *Negative whole numbers* represent some type of whole units. To understand negative numbers, it is best to review the time line shown in Chapter 1. Take a look at the following chart.

$$—|—|—|—|—|—|—|—|—|—|—$$
$$-5 \quad -4 \quad -3 \quad -2 \quad -1 \quad 0 \quad +1 \quad +2 \quad +3 \quad +4 \quad +5$$

Negative Integers   Positive Integers

Note that the numbers become larger as you move from the zero to the right. Also notice that they become smaller as you move from the zero to the left. Picture the zero as if it were a large plate glass window, with the right side indicating your positive net worth and the left side indicating your negative net worth (debt). Imagine that you have five dollars in your bank account but owe three dollars on your credit card. What would be your true net worth? If you took the five dollars from your bank account and paid off the three that you owed on the credit card, how much would you have left? This becomes a simple addition math problem: $5 - 3 = 2$. Later, when we explore positive and negative numbers, the rule to follow is to subtract the smaller number from the larger number, and your answer will be the larger number. With this problem, the 3 is taken from the 5 and the answer would be 2.

This is a simple explanation of adding positive and negative whole numbers. However, it should provide you with sufficient knowledge to proceed to the next step, adding positive and negative integers.

Adding Positive and Negative Integers

Occasionally, addition problems will be encountered requiring an individual to add positive and negative integers. The procedure for solving these problems involves both the addition and the subtraction processes. The general procedure is to:

1. Determine the sum of the positive integers.
2. Determine the sum of the negative integers.
3. Determine the difference between the two by subtracting the smaller number from the larger number. The difference will take the sign of the larger number.

Problem: Solve the following addition problem.

$$\begin{array}{r} -47,235 \\ +76,482 \\ +63,823 \\ -23,843 \end{array}$$

Solution:

1. Determine the sum of the positive integers.

$$\begin{array}{r} +76,482 \\ +63,823 \\ \hline +140,305 \end{array}$$

2. Determine the sum of the negative integers.

$$\begin{array}{r} -47,235 \\ -23,843 \\ \hline -71,078 \end{array}$$

3. Determine the difference between the two answers by subtracting the smaller number from the larger number. In this problem, the positive integers are the largest. The problem would appear as:

$$\begin{array}{r} +140,305 \\ -71,078 \\ \hline 69,227 \end{array}$$

The answer would take the positive sign because the larger sum was the positive numbers.

Answer: +69,227

Note: It is good practice to check the two addition problems prior to starting the subtraction portion.

This problem ended up being both an addition and a subtraction problem rather than a simple addition problem. Consequently, information regarding the method of determining whether or not a subtraction problem has been worked correctly will be delayed until after the chapter on working with subtraction problems.

Problem: Solve the following addition problem.

$$\begin{array}{r} -37,291 \\ +62,351 \\ -75,284 \\ +35,762 \end{array}$$

Solution

1. Determine the sum of the positive integers.

$$+62,351 \\ +35,762 \\ \hline +98,113$$

2. Determine the sum of the negative integers.

$$-37,291 \\ -75,284 \\ \hline -112,575$$

3. Determine the difference between the two numbers by subtracting the smaller number from the larger number. In this problem, the negative integers are the largest. The problem would appear as:

$$-112,575 \\ +98,113 \\ \hline -14,462$$

The answer would take a negative sign because the largest sum was the negative numbers.

Answer: $-14,462$

Practice Adding Positive and Negative Problems

Following are three practice positive and negative addition problems for the reader to solve.

Problem 1: Solve the following addition problem.

$$-78,459 \\ -55,738 \\ +45,893 \\ +64,572$$

Solution

1. Determine the sum of the positive integers.

$$+45,893 \\ +64,572$$

2. Determine the sum of the negative integers.

$$-78,459 \\ -55,738$$

3. Determine the difference between the two numbers by subtracting the smaller number from the larger number. In this problem, the negative integers are larger. Your problem should have appeared as:

$$-134,197 \\ +110,465 \\ \hline -23,732$$

Problem 2: Solve the following addition problem.

$$\begin{array}{r} +56,738 \\ -99,892 \\ +34,770 \\ -32,134 \\ \hline \end{array}$$

Solution:

1. Determine the sum of the positive integers.

$$\begin{array}{r} +56,738 \\ +34,770 \\ \hline \end{array}$$

2. Determine the sum of the negative integers.

$$\begin{array}{r} -99,892 \\ -32,134 \\ \hline \end{array}$$

3. Determine the difference between the two numbers by subtracting the smaller number from the larger number. In this problem, the negative integers are larger. Your problem should have appeared as:

$$\begin{array}{r} -132,026 \\ +91,508 \\ \hline -40,518 \end{array}$$

Problem 3: Solve the following addition problem.

$$\begin{array}{r} +56,829 \\ -87,435 \\ +74,823 \\ +89,472 \\ \hline \end{array}$$

Solution:

1. Determine the sum of the positive integers.

$$\begin{array}{r} +56,829 \\ +74,823 \\ +89,472 \\ \hline \end{array}$$

2. Determine the sum of the negative integers.

$$-87,435$$

3. Determine the difference between the two numbers by subtracting the smaller number from the larger number. In this problem, the positive integers are larger. Your problem should have appeared as:

$$\begin{array}{r} +221,124 \\ -87,435 \\ \hline +133,689 \end{array}$$

Sample Word Problems

Some of the key words that indicate that a given problem is an addition problem are: sum of, added to, combined, and total of. These words when changed to a mathematic formula may appear as follows.

Sum of

Problem: Engine 7, Engine 11, and Truck 4 were working at a fire in the central area of Anytown. Engine 7 was staffed with a crew of 4, Engine 11 was staffed with a crew of 5, and Truck 4 was staffed with a crew of 5. What was the sum of the members working at the fire?

Solution: The number (sum) of the members working at the fire can be determined by adding the number on Engine 7, the number on Engine 11, and the number on Truck 4. Expressed as a number problem, it would appear as:

$$X = 4 + 5 + 5 = 14$$

Added to

Problem: During the last brush fire season in Southern California, at one time three major fires were burning: one in Malibu, one near Lake Arrowhead, and one in the San Diego area. There were 1,252 firefighters working on the Malibu fire, 845 on the Lake Arrowhead fire, and 1,625 on the San Diego fire. What was the total number of firefighters working on the Lake Arrowhead fire added to those working on the Malibu fire?

(***Note:*** The numbers assigned to these fires in this problem are used as a sample. They are not correct.)

Solution: As with all word problems, it is important that the problem be read thoroughly prior to starting on finding a solution. The problem asked for the total number of firefighters working on the Lake Arrowhead fire (845) added to those working on the Malibu fire (1,252). Expressed as a number problem, it would appear as:

$$X = 1252 + 845 = 2097$$

Combined

Problem: The Southern county of the state has three volunteer fire departments: The Pine department has 8 stations and 162 volunteers, The High Woods department has 7 stations and 89 volunteers, and the Plains department has 7 stations and 212 volunteers. What is the combined total of the volunteers on the three departments?

Solution: The problem asks for the combined total of volunteers on the Pine department (162), High Woods department (89), and Plains department (212). Expressed as a number problem, it would appear as:

$$X = 162 + 89 + 212 = 463$$

Total of

Problem: Three major wildfires occurred in Oklahoma last year. A total of 23 homes were lost in the first fire, 42 in the second fire, and 55 in the third fire. What was the total number of homes lost in the three fires?

Solution: The problem asks for the total number of homes lost in the three fires. Expressed as an addition problem, it would appear as:

Total number of homes lost = homes lost in the first fire + homes lost in the second fire + homes lost in the third fire. Expressed as an equation problem, it would appear as:

$$X = 23 + 42 + 55 = 120$$

ADDITIONAL WORD PROBLEMS

Problem: During a three-month period, Engine 6 had three fairly good fires. The fire loss in the first fire was reported to be $76,000; the second fire, $176,735; and the third fire, $88,460. What was the total loss in the three fires?

Solution: This problem is an addition problem requiring the reader to add the following three fire loss figures.

$$\begin{array}{r} \$\ 76,000 \\ \$176,735 \\ \underline{\$\ 88,460} \end{array}$$

Adding the three figures would produce the following result.

$$\begin{array}{r} \$\ 76,000 \\ \$176,735 \\ \underline{\$\ 88,460} \\ \$341,195 \end{array}$$

Problem: An ambulance company runs out of Engine 4's station. During a six-month period, one of the EMTs worked the following overtime hours:

January — 43; February — 37; March — 39; April — 45; May — 29; June — 18

How many overtime hours did she work during the six-month period?

Solution: The problem is an addition problem requiring the reader to add:

$$43 + 37 + 39 + 45 + 29 + 18$$

Following is the result of the addition.

$$\begin{array}{r} 43 \\ 37 \\ 39 \\ 45 \\ 29 \\ \underline{18} \\ 211 \end{array}$$

Problem: Recently, the local technical school announced a ten-month course in practical nursing. The school emphasized the fact that the health care field is one of the fastest growing occupational areas in the nation today. The cost for a student enrolling in the program includes a $50 application fee and $7,865 for tuition. In addition, books and other fees are $1,221, and parking costs $644. What is the total cost for a student enrolling in the program?

Solution: This addition problem requires the reader to add the following:

Application fee	$ 50
Tution	$7,865
Books and other fees	$1,221
Parking costs	$ 644

Following is the result of the adding.

$$\begin{array}{r} 50 \\ 7{,}865 \\ 1{,}221 \\ \underline{644} \\ \$9{,}780 \end{array}$$

Reader's Practice Problems—2.3

Note: The following problems are not factual, or realistic.

Problem 1:
A strong earthquake occurred in the southern part of the state in May. The epicenter was near the town of Woodberry, which has a population of 23,560. Twenty-five buildings were severely damaged here. In the neighboring city of Boomtown, six buildings were severely damaged and an additional three were severely damaged in another city close by. What was the total number of buildings severely damaged in the three cities combined?

Problem 2:
All firefighters in the community of Rome are required to perform physical fitness training every duty day. During the first five shifts in June, members of Engine 7 devoted 45 minutes to physical training the first shift, 42 minutes the second shift, one hour and ten minutes the third shift, 55 minutes the fourth shift, and 46 minutes the fifth shift. How much time did the members devote to physical training during the first five shifts of the month?

Problem 3:
At a large fire in the community of Britian, Engine 2 laid 850 feet of $2\frac{1}{2}$-inch hose and 150 feet of 2-inch hose. At the same fire, Engine 7 laid 350 feet of $2\frac{1}{2}$-inch hose and 300 feet of $1\frac{1}{2}$-inch hose. Another 700 feet of $2\frac{1}{2}$-inch hose was laid by Engine 9. What was the total amount of $2\frac{1}{2}$-inch hose laid by the three engine companies combined?

The answers are provided in the Reader's Practice Problems section of the Appendix.

REVIEW QUESTIONS

1. What is the definition of an addend?
2. What is the word that describes the result obtained by adding together two or more digits or numbers?
3. What process is used to determine whether or not the answer to an addition problem is correct?
4. What procedure is used to add positive and negative integers?
5. How can the procedure for determining the "excess of nines" of a smaller addition problem be shortened?

Test 2

First, reproduce the addition chart without referring to it. Then check your chart against the chart shown in the earlier part of this session.

1. Work each of the following five problems and check your answer by the "excess of nines" method. If your answer does not check out, do it again until it does check out. When you finish these problems, you should have a 100% success rate. In other words, you should receive a perfect score on these five problems.

a. 4,562	b. 97,510	c. 17,403	d. 33,486	e. 49,063
8,745	64,507	33,425	46,831	66,435
7,754	27,940	65,832	85,394	66,093
8,543	64,835	23,568	12,756	76,945
4,486			76,942	21,635

   34090

2. Add the following. $-4,567 +5,642 +3,475 -5,489$
3. Add the following. $+3,464 +9,823 -21,452 +719$
4. A riot occurred in Newtown that lasted three days. On the first day, Engine Six laid 1,200 feet of $2\frac{1}{2}$-inch hose and 550 feet of $1\frac{1}{2}$-inch hose. On the second day, the company laid exactly the same amount of hose as on the first day. On the third day, they laid 1,150 feet of $2\frac{1}{2}$-inch hose and 350 feet of $1\frac{1}{2}$-inch hose. What was the total amount of hose they laid during the three days? Don't forget to prove that your answer is correct.
5. During a three-month period, Engine 7 had four working fires. At the first fire, they pumped 1,200 gallons of water; at the second, they pumped 950 gallons of water; at the third, 2,500 gallons; and at the fourth, only 300 gallons. What was the total amount of water pumped at the four fires?
6. During the month of May, the Newtown fire department responded to 68 calls for medical assistance. In June, the amount was 92. July was a relatively quiet month with only 44 responses. What was the total number of responses made during the three-month period?

   Starting at this point, many of the problems that will be given in the chapter tests, and at various parts of individual chapters, will be based on fire department formulas. To many of you, working with formulas is as routine as waking up in the morning. However, most of the formulas you have worked with are different than the fire departments' formulas that will be given to you. To others, working with fire department formulas is a new way of life.

   Most of the fire department formulas are taken from a hydraulic course or book where the formula is based upon determining the data needed to solve a hydraulic problem. In this book, where necessary, the formula is given and the data needed to solve the formula is either provided in a word problem, photo, or drawing. Just be patient and you will soon be able to swing into the new procedure with little difficulty.

   The following problems use a combination of letters common to the fire service:

   FL—friction loss in psi (pounds per square inch)

   NP—nozzle pressure (the pressure at the nozzle tip in psi)

   BP—back pressure (the pressure in psi that is working against the pumper when the pumper is supply water to an elevation above the pumper)

Figure 2.1

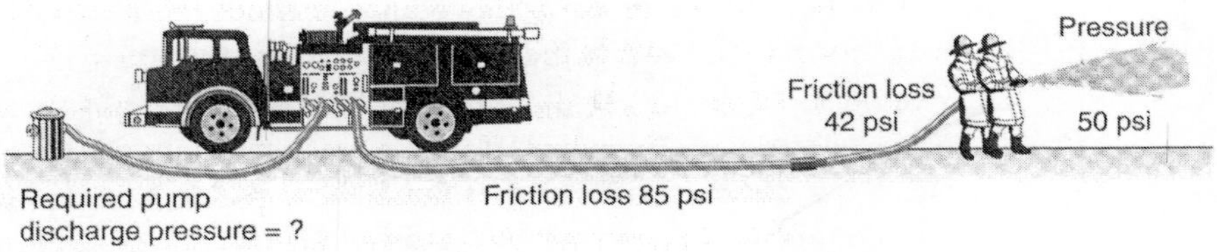

Required pump discharge pressure = ?
Friction loss 85 psi
Friction loss 42 psi
Pressure 50 psi

FLOW—the discharge of water from a nozzle tip or pumper given in gpm (gallons per minute)

RPDP—the required pumper discharge pressure in psi

7. To solve this problem, use the information provided in Figure 2.1.

   Determine the RPDP. The RPDP in this problem equals the total friction loss in the hose line plus the nozzle pressure. It is only necessary to add the two to obtain the RPDP. However, the entire formula for determining the RPDP is expressed as:

   $$RPDP = NP + FL + BP + AFL - FP$$

   where NP = nozzle pressure
   FL = friction loss
   BP = back pressure
   AFL = appliance friction loss
   FP = forward pressure

8. In Figure 2.2, determine the amount of water being discharged from the pumper. The amount being discharged from the pumper is the total amount discharged from the three nozzle tips.

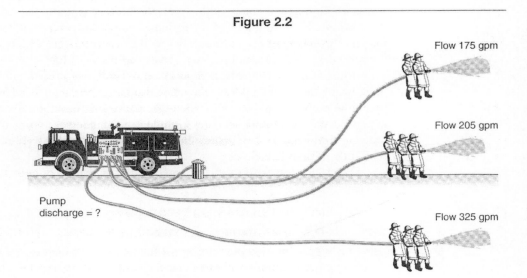

Figure 2.2

Pump discharge = ?
Flow 175 gpm
Flow 205 gpm
Flow 325 gpm

Figure 2.3

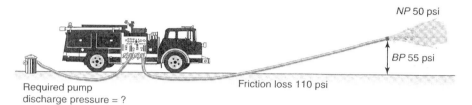

9. In Figure 2.3, determine the RPDP. The entire formula for the RPDP was given in Problem 7. In this problem, the RPDP equals the friction loss in the hose line plus the nozzle pressure plus the back pressure. Just add those three items to solve the problem. However, keep in mind that the entire formula for the RPDP is expressed as:

$$RPDP = NP + FL + BP + AFL - BP$$

10. In Figure 2.4, determine the RPDP. The RPDP in this problem is the total pressure required for the friction loss in the hose lines leading into the

Figure 2.4

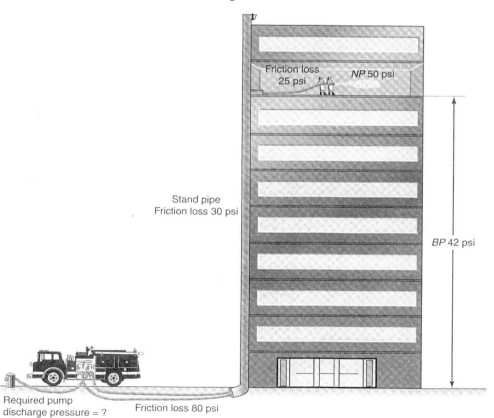

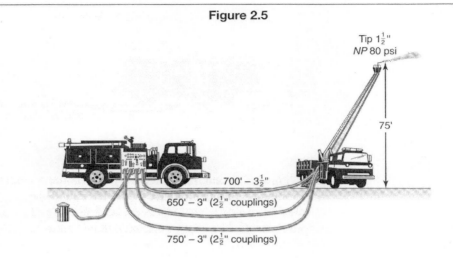

Figure 2.5

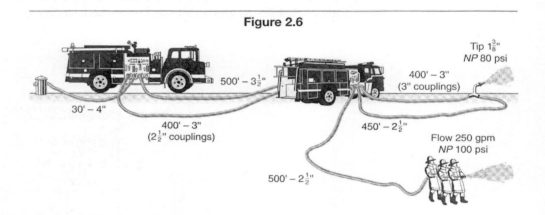

Figure 2.6

standpipe, the friction loss in the hose line being worked on the 9th floor, the nozzle pressure for the nozzle being used on the 9th floor, the friction loss in the standpipe, and the back pressure.

11. In Figure 2.5, determine the total amount of hose laid from the pumper to the ladder pipe.

12. In Figure 2.6, determine the total amount of hose laid out in this situation. It includes the hose laid from the hydrant into the left-hand pumper, the two lines laid from the first pumper into the second pumper, the two lines laid from the second pumper into the portable monitor, and the hose line used by the three firefighters.

13. In Figure 2.7, determine the RPDP (see Problem 7). In this problem, the RPDP equals the friction loss in the hose line, the nozzle pressure on the nozzle tip being used on the fire, and the back pressure.

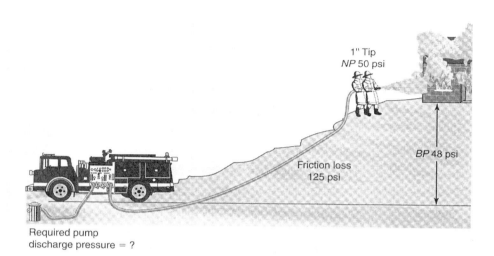

Figure 2.7

14. In Figure 2.8, determine the total amount of hose laid out at this incident. The hose laid out includes the 20-feet suction line from the hydrant into the top pumper, the two lines laid from the top pumper into the second pumper, the two lines laid from the second pumper into the ladder pipe, the suction line laid from the hydrant into the bottom pumper, and the two lines laid from the bottom pumper into the ladder pipe.

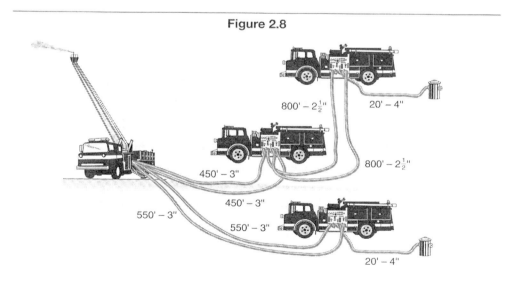

Figure 2.8

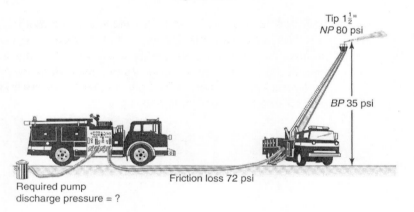

Figure 2.9

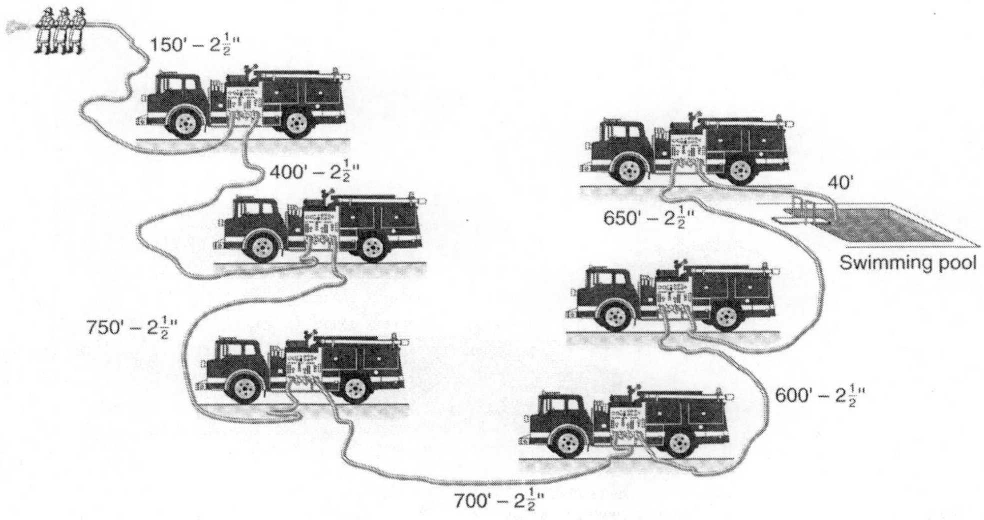

Figure 2.10

15. In Figure 2.9, determine the total amount of hose laid out at this relay operation. It includes the suction line from the swimming pool into the first pumper, all the lines from one pumper to another, and the line from the last pumper that is in operation by the firefighters.

16. In Figure 2.10, determine the RPDP (see Problem 7). The RPDP is the combination of the friction loss in the hose line, the back pressure, and the nozzle pressure on the ladder pipe tip.

17. The task of a pumper is to provide sufficient water at a sufficient pressure to obtain a good discharge flow at the end of all the hose lines taken off the pumper. The pressure required to obtain this goal is referred to as the required pump discharge pressure. The formula for determining this pressure is

$$RPDP = NP + FL + BP + AFL - FP$$

Figure 2.11

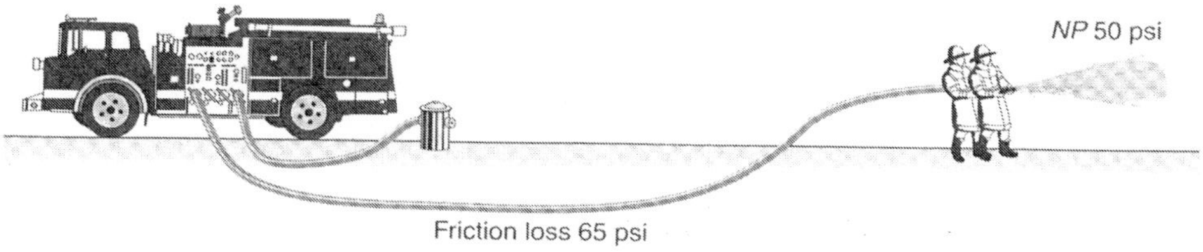

Friction loss 65 psi
NP 50 psi

where NP = nozzle pressure
FL = friction loss in the hose
BP = back pressure
AFL = appliance friction loss
FP = forward pressure

All factors involved with the formula are seldom present in a single problem. Sometimes when hand lines are used, only a pump pressure is required to provide for the nozzle pressure and the friction loss in the hose lines. That is the situation shown in Figure 2.11. What is the pressure requirement for this problem?

18. The situation in Figure 2.12 is quite different. What is the required pump discharge pressure for this layout?

19. Determine the "excess of nines" of the following addition problem by the shortcut method.

$$\begin{array}{r} 764 \\ 516 \\ 238 \end{array}$$

20. Determine the "excess of nines" of the following addition problem by the shortcut method.

$$\begin{array}{r} 4{,}783 \\ 6{,}251 \\ \underline{6{,}743} \end{array}$$

Figure 2.12

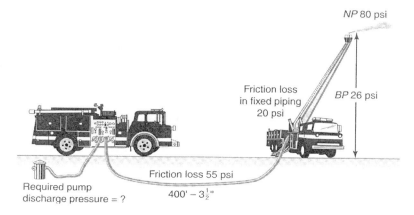

NP 80 psi
Friction loss in fixed piping 20 psi
BP 26 psi
Friction loss 55 psi
Required pump discharge pressure = ?
400' – 3½"

21. Determine the "excess of nines" of the following addition problem by the shortcut method.

    71,462
    83,557
    64,274
    17,842

22. Add

    67,498
    42,753
    83,412
    38,564

23. Add

    +6,784
    +9,456
    −5,834
    +2,376

24. Add

    −4,527
    −5,783
    +3,031
    +9,556

25. At a large fire in the center of town, Engine 5 laid 550 feet of $2\frac{1}{2}$-inch hose and 200 feet of $1\frac{1}{2}$-inch hose. At the same fire, Engine 8 laid 750 feet of $2\frac{1}{2}$-inch hose and 500 feet of 1-inch hose, Engine 7 laid 450 feet of 2-inch hose whereas Engine 10 laid 850 feet of $2\frac{1}{2}$-inch hose. How much $2\frac{1}{2}$-inch hose was laid at this fire?

26. At a small house fire, pumper 33 supplied water through a $2\frac{1}{2}$-inch hose that was laid up a hill to a point 550 feet above the pumper. The nozzle pressure on the line was 50 psi, the friction loss in the line was 130 psi, and the back pressure was 28 psi. What pressure was the RPDP for this line? (See Problem 7.)

27. Three lines were laid to a house fire on 31st street. Engine 5 laid 450 feet of hose, Engine 8 laid 650 feet of hose, and Engine 12 laid 950 feet of hose. How much hose was laid at this fire?

28. During a six-month period, Engine 2 responded to three fires where they laid hose. At the first fire they laid 450 feet of $2\frac{1}{2}$-inch hose and 150 feet of $1\frac{1}{2}$-inch hose, at the second fire they laid 550 feet of $1\frac{1}{2}$-inch hose, and at the third fire they laid 450 feet of $2\frac{1}{2}$-inch hose. What was the total amount of hose laid at the three fires?

29. During a three-month period, Engine 9 responded to four fires and a number of calls for emergency service. In May they responded to 24 calls for emergency service, in June they responded to one fire call and 33 calls for emergency service, and in July they responded to 3 fire calls and 42 calls for emergency service. During the three-month period, to how many calls for emergency service did they respond? (Do not count the fire calls.)

30. The formula for determining the RPDP (required pump discharge pressure) is:

$$RPDP = NP + FL + BP + AFL - FP$$

where   NP = nozzle pressure
   FL = friction loss
   BP = back pressure
   AFL = appliance friction loss
   FP = forward pressure

At a fire on Friday afternoon, Engine 3 laid 400 feet of $2\frac{1}{2}$-inch hose with a nozzle pressure of 50 psi and a friction loss of 110 psi. What was the RPDP for this fire?

31. During a one-week period, Paramedic Smith worked 8 hours on Monday, 8 hours on Tuesday, 9 hours on Wednesday, 9 hours on Thursday, and 8 hours on Friday. What was the total amount of hours she worked during the week?

Chapter Summary

This chapter reviewed the process used for adding whole numbers, and introduced a system for using the "excess of nines" to prove whether or not the answer to a solved addition problem was correct. It defined the two basic words used in an addition problem, and introduced a shortcut method for determining the "excess of nines" of an addition problem. In addition, this chapter provided review questions as well as both basic and practical problems for the reader to solve.

CHAPTER 3
WORKING WITH WHOLE NUMBER MULTIPLICATION PROBLEMS

KEY TERMS

multiplicand product
multiplier squaring

OBJECTIVES AND GOALS

Upon completing this chapter, the reader should be able to:

- Mentally multiply any two single-digit numbers.
- Mentally multiply any single digit by nine.
- Mentally multiply any two-digit number by eleven.
- Mentally multiply two 'teen numbers.
- Mentally square any two-digit number ending in five.

INTRODUCTION

This chapter is both a review chapter and an introduction chapter. Readers are provided with a review of the multiplication process and the times tables that they most likely learned a number of years ago. However, they are also introduced to a number of multiplication shortcuts that may be completely new to them. A particularly valuable one that can be used in day-to-day operations is a method of mentally multiplying 'teen numbers. Review questions and a test are included at the end of the chapter.

WORDS TO REMEMBER

multiplicand—The number being multiplied.

multiplier—The number of times the multiplicand is multiplied.

product—The answer to the multiplication.

The words are illustrated by the following:

$$\begin{array}{r} 5 \\ \times 3 \\ \hline 15 \end{array} \begin{array}{l} - \text{ multiplicand} \\ - \text{ multiplier} \\ - \text{ product} \end{array}$$

A Review of the Multiplication Process

From a practical standpoint, multiplication is nothing more than another method of solving an addition problem. Compare the result of solving each of the following problems.

```
     6 — multiplicand          6 — addend
    ×4 — multiplier            6 — addend
    24 — product               6 — addend
                               6 — addend
                              24 — sum
```

The Multiplication Tables

If your memory takes you back that far, you might remember that much of the math you learned was taught in the elementary grades. Of the different applications to which a youngster was exposed, to the majority, the most demanding and difficult was memorizing the multiplication tables. Of the tables, the nines table was generally considered the most difficult.

Today, because of the wide use of calculators and the limited time a teacher has for fulfilling the requirements in other disciplines, many school districts have lowered the requirement that students memorize the times tables. To an old-timer, this is not acceptable, and it is extremely discouraging to see some high school graduates not even being capable of making change for a dollar. If it weren't for the computers built into cash registers, some individuals would no longer be qualified to hold their present position. For this reason, a multiplication chart and a simple procedure for learning the nines table has been included in this book.

THE MULTIPLICATION CHART

	1	2	3	4	5	6	7	8	9
1	1	2	3	4	5	6	7	8	9
2	2	4	6	8	10	12	14	16	18
3	3	6	9	12	15	18	21	24	27
4	4	8	12	16	20	24	28	32	36
5	5	10	15	20	25	30	35	40	45
6	6	12	18	24	30	36	42	48	54
7	7	14	21	28	35	42	49	56	63
8	8	16	24	32	40	48	56	64	72
9	9	18	27	36	45	54	63	72	81

This chart can be used in one of two ways. A number in the top horizontal column can be multiplied by a number in the left vertical column, or a number in the left vertical column can be multiplied by a number in the top horizontal column. Where the two columns intersect provides the answer. For example, if you want to multiply 7 × 6, select the 7 from the left-hand vertical column and the 6 from the top horizontal column. Note that the two columns intersect at 42, the answer to the multiplication.

A Review of the Times Tables

However, to solve multiplication problems, it is essential that an individual memorize the multiplication tables. Although this took place many years ago in elementary school, it is worthwhile to review it prior to tackling the multiplication process.

Two's Tables	Three's Tables	Four's Tables
2 × 2 = 4	3 × 2 = 6	4 × 2 = 8
2 × 3 = 6	3 × 3 = 9	4 × 3 = 12
2 × 4 = 8	3 × 4 = 12	4 × 4 = 16
2 × 5 = 10	3 × 5 = 15	4 × 5 = 20
2 × 6 = 12	3 × 6 = 18	4 × 6 = 24
2 × 7 = 14	3 × 7 = 21	4 × 7 = 28
2 × 8 = 16	3 × 8 = 24	4 × 8 = 32
2 × 9 = 18	3 × 9 = 27	4 × 9 = 36

Five's Tables	Six's Tables	Seven's Tables
5 × 2 = 10	6 × 2 = 12	7 × 2 = 14
5 × 3 = 15	6 × 3 = 18	7 × 3 = 21
5 × 4 = 20	6 × 4 = 24	7 × 4 = 28
5 × 5 = 25	6 × 5 = 30	7 × 5 = 35
5 × 6 = 30	6 × 6 = 36	7 × 6 = 42
5 × 7 = 35	6 × 7 = 42	7 × 7 = 49
5 × 8 = 40	6 × 8 = 48	7 × 8 = 56
5 × 9 = 45	6 × 9 = 54	7 × 9 = 63

Eight's Tables	Nine's Tables
8 × 2 = 16	9 × 2 = 18
8 × 3 = 24	9 × 3 = 27
8 × 4 = 32	9 × 4 = 36
8 × 5 = 40	9 × 5 = 45
8 × 6 = 48	9 × 6 = 54
8 × 7 = 56	9 × 7 = 63
8 × 8 = 64	9 × 8 = 72
8 × 9 = 72	9 × 9 = 81

Multiplying by Nine

When memorizing the multiplication tables, it is valuable to be aware of a simple method of mentally multiplying a single digit by nine. You might appreciate having a couple of easy methods. Examine the following.

Take one less than the digit being multiplied by nine and add to it whatever it takes to total nine. For example:

Problem: Multiply 9 × 4 = ?

    One less than 4 is 3.

    Take what has to be added to the 3 to make 9. It takes 6.

    3 + 6 = 9

    Combine the 3 and the 6.

    Therefore, 9 × 4 = 36

Another example:

Problem: Multiply 9 × 8

    One less than 8 is 7

    Take what has to be added to the 7 to make 9. It takes 2.

Figure 3.1

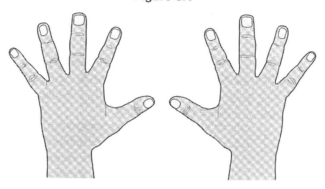

Combine the 7 and 2.

$7 + 2 = 9$

Therefore, $9 \times 8 = 72$

There is a manual method that is even easier. Try the following:

1. Hold both your hands in front of you with the fingers spread out (see Figure 3.1).
2. Select a number you wish to multiply by nine and bend that finger down. For example, if you wishs to multiply 9×4, hold down the fourth finger from the left (see Figure 3.2).
3. Count the fingers to the left of the one bent down (there are 3). Then count the number to the right of the finger held down (there are 6).
4. Therefore, $9 \times 4 = 36$

There is also an interesting fact regarding the nines table. Take a look at the following:

$$9 \times 2 = 18$$
$$9 \times 3 = 27$$
$$9 \times 4 = 36$$
$$9 \times 5 = 45$$
$$9 \times 6 = 54$$
$$9 \times 7 = 63$$
$$9 \times 8 = 72$$
$$9 \times 9 = 81$$

Figure 3.2

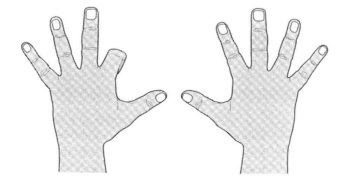

Notice that the two digits in the product for each number multiplied totals 9. This phenomenon was used to design the following problem. If you have already been exposed to it, skip it. However, if you have not, then take it and complete it as if it were given to you verbally. See what answers you obtain.

1. Think of a number from two to nine.
2. Multiply the number by 9.
3. You now have a two-digit number.
4. Add the two digits.
5. You now have a single digit number.
6. Subtract five from this number. You still have a single digit number.
7. Using one as A, two as B, three as C, and so forth, convert your number to a letter.
8. Think of a country that starts with that letter.
9. Take the last letter of that country and think of an animal that starts with that letter.

The country you thought of was Denmark and the animal you thought of was a kangaroo. If this is true, you are probably wondering how the person who presented you with the problem was able to do it. The answer is with the nines multiplication table.

Notice that regardless of the number you thought of from two to nine, the two digits of the number you obtained by multiplying it by nine totaled nine when they were added. You were then instructed to subtract five from the nine. This left you with a four. You were then instructed to convert the four to a letter. The letter you obtained by the conversion was a D. You were then asked to think of a country that started with D. There are very few of them other than Denmark that a normal person generally knows. Can you think of one? You were asked to think of an animal starting with the last letter of the country you selected. The last letter of Denmark is a K. There are also very few animals that have a name starting with a K. Can you think of any other than a koala bear?

Solving a Multiplication Problem

A multiplication problem may be presented as: Multiply 3 times 5; 3 $\times$ 5; 3 · 5; (3)(5); or 5
$\times$ 3

Of the methods indicated, the use of an $\times$ is probably the most common.

Take a look at the following multiplication problem. No one who knows his or her multiplication tables would have any trouble solving it. The solution would appear as follows:

$$\begin{array}{r} 4{,}273 \\ \times \quad 5 \\ \hline 21{,}365 \end{array}$$

However, the individual who solved it would probably cringe when asked to solve the following problem.

$$\begin{array}{r} 4{,}273 \\ \times \ 3{,}425 \\ \hline \end{array}$$

Although this problem may appear to be difficult, in reality it consists of solving four small multiplication problems and one rather easy addition problem. The multiplier for each of the four individual multiplication problems can be identified by their column location in the original problem. These locations are illustrated in the following:

```
4  2  7  3   — multiplicand
3  4  2  5   — multiplier
t  h  t  o
h  u  e  n
o  n  n  e
u  d  s  s
s  r
a  e
n  d
d  s
s
```

The following illustrates the four multiplication problems required to be solved. The multiplier of each of the problems can be identified as to its column location in the original multiplier as illustrated above.

thousands multiplier	hundreds multiplier	tens multiplier	ones multiplier
4,273 × 3	4,273 × 4	4,273 × 2	4,273 × 5

Their solution is rather simple. Take a look.

4,273 × 3	4,273 × 4	4,273 × 2	4,273 × 5
12,819	17,092	8,546	21,365

The next step is where some individuals make a mistake. This is placing the product of each of the individual problems in the wrong place. The principle to remember is that the right-hand digit of the product produced by a multiplier should be placed in the same column as the multiplier. The first product to be placed in the answer space is the ones product, the second the tens product, the third the hundreds product, and the fourth the thousands product. This principle is illustrated in the following:

```
      4  2  7  3
      3  4  2  5
   2  1  3  6  5   ← ones multiplier product
   8  5  4  6      ← tens multiplier product
1  7  0  9  2      ← hundreds multiplier product
1  2  8  1  9      ← thousands multiplier product

t  h  t  o
h  u  e  n
o  n  n  e
u  d  s  s
s  r
a  e
n  d
d  s
s
```

All that is left is to solve the addition of the products of the four individual multiplication problems. The solution is illustrated in the following.

```
            4   2   7   3
            3   4   2   5
        2   1   3   6   5
        8   5   4   6
    1   7   0   9   2
1   2   8   1   9
1   4   6   3   5   0   2   5     ← problem product
```

It is now time to prove that the problem has been worked correctly. However, before proceeding in that direction, it is time to bring the reader into the multiplication process. If you remember all of the multiplication tables, you should not have any trouble working all of the following problems. However, if you do have trouble working any of them, you should go back and rememorize the multiplication tables. Failing to do so will make the completing of the book extremely difficult.

Problems: One-Digit Multiplication

Work each of the following problems as quickly as you can, and place your answer directly under the problem. When you finish, go back to the multiplication tables and check to see if *all* of your answers were correct. If they were not, you should take time to make sure you know the tables.

2	7	5	6	9	4	6	7	3	8	2	6	7
7	6	9	4	5	3	6	9	2	4	6	4	7

5	8	3	7	9	3	5	2	8	8	5	4	2
7	4	5	7	4	2	4	8	6	3	3	8	7

8	3	8	4	7	3	2	8	9	5	4	2	4
7	6	3	4	7	5	9	3	5	3	8	2	5

$3 \times 7 =$ $9 \times 4 =$ $7 \times 8 =$ $5 \times 4 =$ $8 \times 2 =$ $8 \times 6 =$

$9 \times 7 =$ $7 \times 6 =$ $2 \times 9 =$ $6 \times 3 =$ $7 \times 6 =$ $5 \times 5 =$

$2 \times 4 =$ $4 \times 8 =$ $6 \times 3 =$ $7 \times 8 =$ $3 \times 3 =$ $4 \times 2 =$

Problems: Two-Digit Multiplication

You should make sure that you know the multiplication tables prior to tackling these problems. When you feel that you can solve them, start working them as quickly as you can. After finishing, check each problem to make sure that you solved it correctly.

67	56	93	37	23	91	78	54	36
34	85	42	82	87	36	53	88	17

13	69	62	68	39	74	23	89	52
34	27	53	31	84	58	76	23	44

74	34	13	64	73	43	72	59	77
66	48	25	33	84	63	31	37	44

Problems: Mix Whole Digits Multiplication

89	6	3×4	54	13	4×63	78	3×34
3	17		6	22		6	

$3 \times 47 =$ $23 \times 7 =$ $43 \times 31 =$ $44 \times 6 =$ $62 \times 23 =$

17	5	83	39	2	19	39	6	34
3	9	11	4	6	78	43	17	22

Reader's Practice Problems—3.1

All of the following are multiplication problems. Solve each of them.

Single-Digit Multipliers

6	8	4	4	6	9	3	5	25	78	35	98	46	54	73	26
2	3	4	5	6	7	8	9	2	3	4	5	6	7	8	9

Two-Digit Multipliers

28	65	87	75	48	92	68	24	67	77	42
53	67	63	84	23	62	52	31	48	83	76

674	937	372	667	438	966	437	547	231
77	73	28	98	63	58	41	32	45

Three-Digit Multipliers

123	456	789	362
563	945	264	693

The answers are found in the Reader's Practice Problems of the Appendix.

Proving a Multiplication Problem

With small multiplication problems, it is easier to determine if the problem has been solved correctly by using the conventional method. With the conventional method, you can divide the product of the multiplication by the divider. If the result is equal to the multiplicand, the problem has been worked correctly. For example, if the problem were to multiply 3 × 249, the result would be:

$$
\begin{array}{r}
249 \\
\times\ 3 \\
\hline
747
\end{array}
$$

To determine if the problem had been worked correctly, divide the product (747) by the divisor (3). The answer would be 249. If the answer is equal to the dividend, the problem has been worked correctly.

However, look at the previous problem that was worked. It was

multiply: 4273 × 3425

Referring back, the product of this multiplication was: 14,635,025. This makes it a little difficult to prove that the multiplication had been correct by using the conventional method of proving the problem. However, let's take a look at proving it by the "excess of nines" method. The multiplication of the problem looked like this:

```
            4 2 7 3
            3 4 2 5
          ─────────
          2 1 3 6 5
          8 5 4 6
        1 7 0 9 2
      1 2 8 1 9
      ─────────────
      1 4 6 3 5 0 2 5   ← problem product
```

You can imagine how difficult this would have been to prove using the conventional method. However, using the "excess of nines" method it would have been different. The problem has been set up as it was originally solved.

Because this was a multiplication problem, the first step is to determine the "excess of nines" of both the multiplicand and the multiplier. The "excess of nines" of the multiplicand is 7 whereas the "excess of nines" of the multiplier is 5.

The next step is to multiply the two "excess of nines": $5 \times 7 = 35 = 8$ ← ← ← ←

The following step is to determine the "excess of nines" of the product.
The product is 14635025.

Let's determine it by adding all the digits.

$1 + 4 + 6 + 3 + 5 + 0 + 2 + 5 = 26 = 8$ ← ← ← ← same ← ← ← ←

Because the "excess of nines" of the multiplicand and product are the same, the problem can be considered as having been worked correctly.

Reader's Practice Problems—3.2

All of the following are multiplication problems. Check the answer by the conventional procedure or by the "excess of nines" method. If it has been worked correctly, move on to the next problem. If it is wrong, correct it so that it checks out to be correct.

Two-digit multipliers

23	76	92	76	45	88	425
63	54	69	84	62	43	79
1,449	4,104	6,348	6,384	2,790	3,784	33,575

933	638	297	556
41	77	38	57
38,253	9,126	11,286	31,692

Three-digit multipliers

365	563	882	976	4523	5674
622	231	589	849	572	327
22,7030	130,053	519,498	828,624	2,587,156	1,855,398

The answers are provided in the Reader's Practice Problems section of the Appendix.

Shortcut Multiplication Procedures

As mentioned in the chapter's introduction, this chapter is both a review chapter and one that introduces some new methods of solving certain types of multiplication problems. The methods that are introduced are referred to as shortcut multiplication procedures.

Five shortcut multiplication procedures are available that are relatively simple to understand and use:

1. Multiplying a two-digit number by eleven.
2. Multiplying two 'teen numbers.
3. Squaring a two-digit number that ends in five.
4. Multiplying a number by ten.
5. Multiplying a number by one hundred.

Multiplying a Two-Digit Number by Eleven

Of all the shortcuts in multiplication, the easiest to work is multiplying a two-digit number by eleven. If the two digits total nine or less, separate the two digits, add them, and then place the result in the center.

Problem: Multiply 11 × 45

Solution: Separate the 4 and the 5   4   5
           Add the two digits   4 + 5 = 9
           Place the result in the center of the two numbers   4   9   5
           Therefore, 11 × 45 = 495

Problem: Multiply 11 × 43

Solution: Separate the 4 and the 3   4   3
           Add the two digits   4 + 3 = 7
           Place the result in the center of the two numbers   4   7   3
           Therefore, 11 × 43 = 473

Problem: Multiply 11 × 27

Solution: Separate the 2 and 7   2   7
           Add the two digits   2 + 7 = 9
           Place the result in the center of the two digits   2   9   7
           Therefore, 11 × 27 = 297

When multiplying a two-digit number by 11, if the two digits total 10 or more, separate the two digits, add them, place the right-hand digit of the result in the center and increase the left-hand digit by one.

Problem: Multiply 11 × 67

Solution: Separate the 6 and the 7   6   7
           Add the two digits   6 + 7 = 13
           Place the right-hand digit of the result in the center   6   3   7
           Increase the left-hand digit (6) by one   7   3   7
           Therefore, 11 × 67 = 737

Problem: Multiply 11 × 89

Solution: Separate the 8 and 9   8   9
           Add the two digits   8 + 9 = 17
           Place the right-hand digit (7) of the result in the center   8   7   9
           Increase the left-hand digit (8) by one   9   7   9
           Therefore, 11 × 89 = 979

Mentally Multiplying Two 'Teen Numbers

When you were in elementary school and forced to memorize the multiplication tables, no one thought of teaching you a method for multiplying two 'teen numbers. The process is relatively easy and can be quite valuable. Following is an example of how it works.

Problem: Multiply 16 × 17

Solution:

1. Select one of the numbers and add the right-hand digit of the other number.

    Either   16 + 7 = 23   or   17 + 6 = 23

2. Add a zero to the end of the result   230
3. Multiply the two right-hand digits of the 'teen numbers

    6 × 7 = 42

4. Add this to the 230

    230 + 42 = 272

5. Therefore, 16 × 17 = 272

Try another one:

Problem: Multiply 13 × 19

Solution:

1. Select one of the numbers and add the right-hand digit of the other number.

    Either   13 + 9 = 22   or   19 + 3 = 22

2. Add a zero to the end of the result   220
3. Multiply the two right-hand digits of the 'teen numbers

    3 × 9 = 27

4. Add this to the 220

    220 + 27 = 247

5. Therefore, 13 × 19 = 247

One more:

Problem: Multiply 18 × 19

Solution:

1. Select one of the numbers and add the right-hand digit of the other number.

    Either   18 plus 9 = 27   or   19 plus 8 = 27

2. Add a zero to the end of the result   270
3. Multiply the two right-hand digits of the 'teen numbers

    8 × 9 = 72

4. Add this to the 270

    270 + 72 = 342

5. Therefore, 18 × 19 = 342

That's how to do it. Remember, it is designed to multiply two 'teen numbers mentally, not by the use of a pencil and paper. You should find the process quite valuable, and you'll be surprised how amazed people are when you use it.

Squaring a Number Ending in Five

Squaring a number refers to multiplying a number by itself. For example, to square 4, multiply 4×4. To square 7, multiply 7×7. Squaring is expressed by placing an elevated 2 to the right and slightly above the number or letter to be squared. For example,

$$4^2 = 4 \times 4 = 16. \quad \text{And} \quad 7^2 = 7 \times 7 = 49$$

This section explains a simple method of squaring a two-digit number ending in five. The procedure to use is:

1. Determine the next larger number ending in five that you wish to square.
   For example, if you wish to square 35, the next larger number ending in five is 45.
2. Take the left-hand digit of this number and multiply it by the left-hand digit of the number you wish to square. In the example, take the 4 and multiply it by 3. The result is $3 \times 4 = 12$.
3. Place a 25 at the end of the product obtained: 1,225

Therefore, $35 \times 35 = 1,225$
Try another one.

Problem: Square 65.

Solution: The next larger number ending in five is 75.

Multiply the 7 in the 75 by the 6 in the 65.

$$6 \times 7 = 42$$

Tack a 25 on the end: 4,225

Therefore, 65 squared, or $65 \times 65 = 4,225$

Try one more.

Problem: Square 75.

Solution: The next larger number ending in five is 85.

Multiply the 8 in the 85 by the 7 in the 75.

$$7 \times 8 = 56$$

Tack a 25 on the end: 5,625

Therefore, 75 squared is 5,625

Multiplying a Number by Ten

The system for multiplying a number by ten is even easier than multiplying a two-digit number by eleven. However, the procedure for multiplying a number by ten is not limited to a two-digit number; it can be used on any number. Simply stated:

To multiply a number by ten, move the decimal point in the number one place to the right.

Problem: Multiply 37 by 10

Solution: Move the decimal point one place to the right: $37 \times 10 = 370$

Problem: Multiply 67.843 by 10

Solution: Move the decimal point one place to the right.

$$67.843 \times 10 = 678.43$$

Multiplying a Number by 100

To multiply a number by 100, move the decimal point two places to the right.

Problem: Multiply 234 by 100

Solution: Move the decimal point two places to the right: 234 = 23,400
Moving the decimal two places to the right = 23,400

Problem: Multiply 874.5682 by 100

Solution: Move the decimal point two places to the right.

$$874.5682 \times 100 = 87,456.82$$

Another thing about multiplying by ten is that it is easy to just multiply the numbers and add the zeros. For example, notice the following problem.

Problem: Multiply 20×30

Solution: Notice that both of the numbers to be multiplied end in zero. Together, there are two zeros. Ignore the two zeros and multiply the 2×3, and then add the zeros on the end of your response. The result would look like the following.

Problem: Multiply 20×30

Solution: Instead of:

$$\begin{array}{r} 20 \\ \times 30 \\ \hline 00 \\ 60 \\ \hline 600 \end{array}$$

Just multiply $2 \times 3 = 6$ and add the two zeros 600

Using this procedure is even more valuable when multiplying larger numbers. Take a look at the following problem.

Problem: Multiply $3,000 \times 1,400$

Solution: With the normal procedure, it would appear as:

$$\begin{array}{r} 3,000 \\ \times 1,400 \\ \hline 0000 \\ 0000 \\ 12,000 \\ 3,000 \\ \hline 4,200,000 \end{array}$$

Using the shortcut procedure, it would appear as follows.

Solution: Multiply $14 \times 3 = 42$

Because there are a total of five zeros, add them to the 42: 4,200,000

Reader's Practice Problems—3.3

All of the following are multiplication problems.

Section 1. Solve all problems mentally.

$9 \times 4 \quad 9 \times 9 \quad 9 \times 5 \quad 9 \times 8 \quad 9 \times 3 \quad 9 \times 6 \quad 9 \times 7$

Section 2. Solve all problems mentally.

$11 \times 43 \quad 11 \times 71 \quad 11 \times 26 \quad 11 \times 54$
$11 \times 67 \quad 11 \times 92 \quad 11 \times 38 \quad 11 \times 57$

Section 3. Solve all problems mentally.

$12 \times 15 \quad 16 \times 17 \quad 13 \times 18 \quad 14 \times 14$

Section 4. Solve all problems mentally.

$25^2 \quad 45^2 \quad 75^2 \quad 85^2$

Section 5. Solve all problems mentally.

$25 \times 10 \quad 56.73 \times 10 \quad 930.56 \times 10 \quad 645 \times 10$
$864.58 \times 100 \quad 95.234 \times 100 \quad 8703.265 \times 100$

The answers are in the Reader's Practice Problems section of the Appendix.

The next section involves multiplying positive and negative integers. However, prior to examining this section, it is probably fitting to see if you understood the shortcut multiplication methods just introduced. Solve each of the following problems. If you have trouble with any of them, go back to where they were introduced and review the solving process. Be extremely careful that you know how to multiply two 'teen numbers. You will most likely find more use for this procedure than any of the others introduced.

Problem 1. Multiply 11×44

Problem 2. Multiply 11×75

Problem 3. Multiply 13×17

Problem 4. Multiply 16×18

Problem 5. Square 55

Problem 6. Multiply 10×365

Problem 7. Multiply 130×700

Multiplying Positive and Negative Integers

The only difference between multiplying positive whole number integers and multiplying positive and negative integers is in the sign of the product.

The rules to follow are:

1. If the sign of the multiplicand and the multiplier are different, the sign of the product will be a minus (negative). For example:

Single-Digit Multiplication

-3	$+9$	$+6$	-2	-4	-5	-8	$+7$
$+8$	-5	-8	$+7$	$+9$	$+2$	$+4$	-2
-24	-45	-48	-14	-36	-10	-32	-14

Double-Digit Multiplication

```
 -43    +32    +83    -37    +82    +58    -35    -92
  +6     -9     -5     +2    -23    -21    +48    +42
 ----   ----   ----   ----   ----   ----   ----   ----
-258   -288   -415    -74    246     58    280    184
                             164    116    140    368
                           -1,886 -1,218 -1,680 -3,864
```

2. If the sign of the multiplicand and the multiplier are the same, the sign of the product will be a plus (positive). For example:

Single-Digit Multiplication

```
 +3    -9    -6    +2    +4    -5    +8    -7
 +8    -5    -8    +7    +9    -2    +4    -2
----  ----  ----  ----  ----  ----  ----  ----
+24   +45   +48   +14   +36   +10   +32   +14
```

Double-Digit Multiplication

```
 +43    -32    +83    -37    +82    -58    +35    -92
  +6     -9     +5     -2    +23    -21    +48    -42
 ----   ----   ----   ----   ----   ----   ----   ----
+258   +288   +415    +74    246     58    280    184
                             164    116    140    368
                           +1,886 +1,218 +1,680 +3,864
```

Reader's Practice Problems—3.4

All of the following problems are multiplication problems.

1. Single-Digit Multiplication Problems

```
 +5    -7    +3    -9    -8    +2    +4
 -7    -2    -8    +9    -5    -6    +3
```

2. Two-Digit Multiplication Problems

```
-77   +32   +74   -39   +47   +32
 +5    +8    -8    +2    -6    -3

-17   +45   +37   -78   -84   +55
+23   +34   -38   +14   +44   +53
```

3. $(-7) \times (+8) =$ $(+9) \times (+7) =$ $(-3) \times (-4) =$
   $(+23) \times (-17) =$ $(-45) \times (-4) =$

The answers are provided in the Reader's Practice Problems section of the Appendix.

REVIEW QUESTIONS

1. What is the definition of product?
2. What is the definition of multiplicand?
3. What is the definition of multiplier?
4. What is the shortcut method for mentally multiplying a single-digit number by nine?
5. What is the shortcut method for mentally multiplying two 'teen numbers?

6. What is the process for mentally multiplying a two-digit number by eleven when the two digits total nine or less?
7. What is the process for mentally squaring any two-digit number ending in five?
8. What is an interesting fact regarding the nines multiplication table?
9. What is meant by "squaring" a number?
10. What sign does the answer take when multiplying two negative numbers?
11. What sign does the answer take when multiplying a negative number by a positive number?
12. Explain how to multiply a number by 10.
13. Explain how to multiply a number by 100.
14. Explain how to mentally multiply a two-digit number by eleven if the two digits total ten or more.

Test 3

1. Fill in the blanks for the following multiplication chart.

	2	3	4	5	6	7	8	9
2								
3								
4								
5								
6								
7								
8								
9								

Solve each of the following.

2. Multiply 435×657
3. Multiply $9,648 \times 3,692$
4. Mentally multiply 11×62
5. Mentally multiply 11×27
6. Mentally multiply 11×85
7. Mentally multiply 13×16
8. Mentally multiply 17×18
9. Mentally square 75
10. Mentally square 55

All or part of the information needed for solving the following problems are contained in the figures that are provided with each problem. The figures are either a drawing or a photo. The formulas for solving the problems are all fire service formulas used either in classroom situations or in the field. If needed, the formula is provided with the problem.

Figure 3.3

Courtesy of Rick McClure, LAFD

11. The aircraft shown in Figure 3.3 carries 3,200 gallons in its drop tank. What is the total amount of water it dropped if it completes 12 drops?
   SOLUTION HELP—Multiply 12 × 3,200

12. The formula for determining the volume of a rectangular container is:

   $$V = lwh$$

   where   V = volume in cubic feet
            l = length in feet
            w = width in feet
            h = height in feet

   How many cubic feet are in the rectangular container shown in Figure 3.4?
   SOLUTION HELP—Multiply 22 × 8 × 10

Figure 3.4

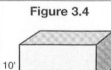

13. There are 1,728 cubic inches in a cubic feet. How many cubic inches are there in the rectangular container shown in Figure 3.5?
   SOLUTION HELP—Multiply 240 × 1,728

Figure 3.5

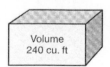

14. Discharge of water from a nozzle tip is expressed in gpm (gallons per minute). What is the total amount of water discharged from the three tips shown in Figure 3.6 in 14 minutes?
   SOLUTION HELP—Add 200 + 175 + 190
   Now multiply 14 × (200 + 175 + 190)

15. The force on a surface is determined by the formula

   $$F = p \times a$$

   where   F = force in pounds
            p = pressure in psi (pounds per square inch)
            a = area in square inches

Figure 3.6

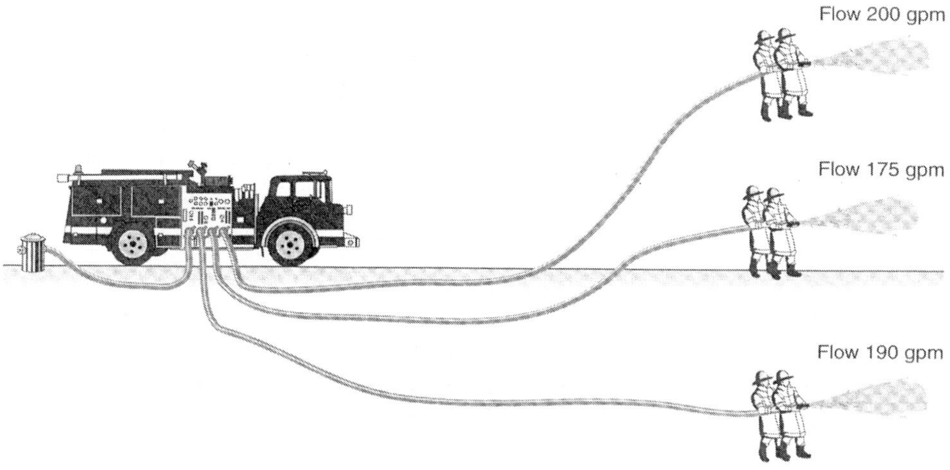

Figure 3.7

How much force is exerted on the clapper valve shown in Figure 3.7?
SOLUTION HELP—Multiply 35 × 99

16. How many gallons of water will be discharged from the tip shown in Figure 3.8 in 1 hour and 20 minutes?
SOLUTION HELP—Multiply 80 × 850

Figure 3.8

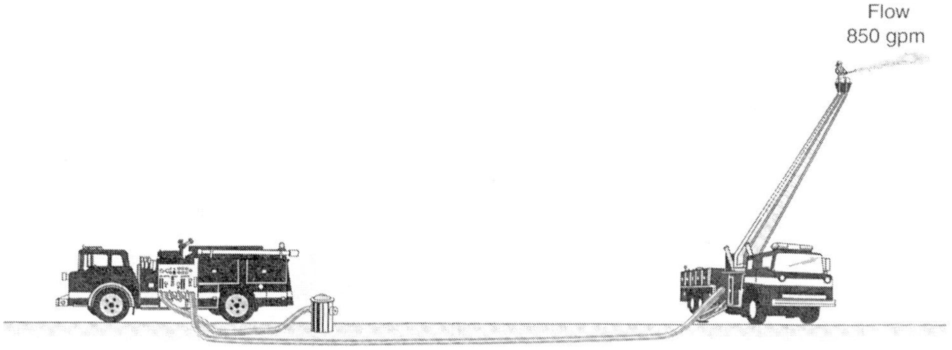

Figure 3.9

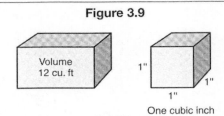

Figure 3.10

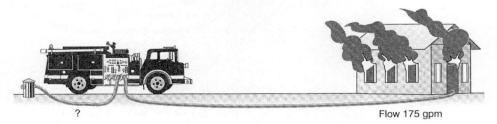

17. There are 1,728 cubic inches in a cubic feet. How many cubic inches are there in the rectangular container shown in Figure 3.9?

    SOLUTION HELP—Multiply 12 × 1728

18. How much water will be discharged on the house fire shown in Figure 3.10 in 17 minutes?

    SOLUTION HELP—17 × 175

19. A section of $2\frac{1}{2}$-inch hose is 50 feet in length. What is the length of the $2\frac{1}{2}$-inch line being used in Figure 3.11?

    SOLUTION HELP—Multiply 18 × 50

20. The nozzle being used by the three firefighters shown in Figure 3.12 is discharging 210 gpm (gallons per minute). How much water will it discharge in one hour and eight minutes? (*Note*: This picture was taken before the implementation of the rule that breathing apparatus would be used in situations like this.)

    SOLUTION HELP—Multiply 68 × 210

21. The Topnotch Fire Department responded to an average of three vehicle fires similar to the one shown in Figure 3.13 every month. How many fires of this type did the department respond to in 17 months?

    SOLUTION HELP—Multiply 17 × 3

22. Multiply +13 × −142

23. Multiply +132 × +34

Figure 3.11

Figure 3.12

Courtesy of Rick McClure, LAFD

24. Multiply -98×-67
25. Multiply $-45 \times +72$
26. While taking a patient's pulse, a paramedic counts 20 beats in 15 seconds. What is the patient's pulse in beats per minute (bpm)?
27. Mentally multiply 87 by 10
28. Mentally multiply 456.79 by 10
29. Mentally multiply 14 by 100
30. Mentally multiply 956.7342 by 100

Figure 3.13

Courtesy of Rick McClure, LAFD

31. Multiply 437
             × 573

32. Mentally multiply 11 × 94

33. Mentally multiply 14 × 16

34. Mentally multiply 15 × 18

35. Mentally square 65

36. How many cubic feet are there in a rectangular container that measures 9 feet by 6 feet by 12 feet?

    SOLUTION HELP—Multiply 9 × 6 × 12

37. There are 1,728 cubic inches in a cubic feet. How many cubic inches are in the rectangular container in Problem 36?

    SOLUTION HELP—Multiply 648 × 1728

38. A $1\frac{1}{4}$-inch tip discharges approximately 325 gpm with a nozzle pressure of 50 psi. How much water will this tip discharge in 75 minutes?

    SOLUTION HELP—Multiply 75 × 325

39. A $1\frac{1}{8}$-inch tip being held by three firefighters is discharging 210 gpm. How much will be discharged by this tip in 1 hour 12 minutes?

    SOLUTION HELP—Multiply 72 times 210.

40. Three engine companies responded to a house fire at 1704 East Main Street. Each of the companies laid 150 feet of $1\frac{1}{2}$-inch hose for use on the fire. Each of the lines discharged 85 gpm. The fire was out in 17 minutes. How much water was used on the fire?

    SOLUTION HELP—The three lines discharged 3 × 85 or 255 gpm. Multiply 17 × 255.

41. Multiply −17 × +65

42. Multiply −46 × −65

43. Mentally multiply 88 by 10

44. Mentally multiply 246.736 by 100

45. Mentally multiply 34.3467 by 1000

Chapter Summary

This chapter reviewed the method for multiplying whole numbers. It also included a review of the multiplication table and introduced the system for checking a solved multiplication problem to determine whether the answer was correct. In addition, shortcut methods were presented for mentally multiplying any single digit by nine, multiplying any two-digit number by eleven, multiplying any two-digit 'teen numbers, mentally squaring any two-digit number ending in five, multiplying a number by ten, and multiplying a number by 100. Multiplying positive and negative integers was also introduced. This chapter also included review questions, as well as a test section of both traditional and practical problems for the reader to solve to determine his or her understanding of the material presented.

CHAPTER 4

SOLVING WHOLE NUMBER DIVISION, SUBTRACTION, AND SQUARE ROOT PROBLEMS

KEY TERMS

difference
dividend
divisor
exponent
minuend
quotient
radical symbol
subtrahend

OBJECTIVES AND GOALS

Upon completing this chapter, the reader should be able to:

- Solve subtraction, division, and square root problems.
- Define dividend, divisor, quotient, minuend, subtrahend, and difference.
- Explain a radical symbol and an exponent.

INTRODUCTION

This chapter is divided into three units: Unit 1 is devoted to the division of whole numbers, Unit 2 concentrates on the subtraction of whole numbers, and Unit 3 focuses on working square root problems. Each unit provides the definition of words used in the process, and reviews the procedure for solving the problem. Each unit also includes both standard and word problems for the reader to solve as well as review questions. Upon completing this chapter, the reader should be able to solve whole number problems involving division, subtraction, and square root.

UNIT 1—DIVISION PROBLEMS

WORDS TO REMEMBER

dividend—The number to be divided.

divisor—The number that will be used to divide the dividend into equal parts.

quotient—The answer to the division problem.

A Review of the Division Process

A division problem is a process of determining how many times one number is contained in another number. Expressed in terms of the words used for division problems, it is a process for determining how many times the divisor is contained in the dividend. The setup for a division problem can be illustrated as follows:

$$\text{divisor} \rightarrow 3\overline{)9} \begin{array}{l} \leftarrow \text{quotient} \\ \leftarrow \text{dividend} \end{array}$$

Instructions for solving division problems are also given in words such as: divide 14 by 7, or by the use of the division symbol ÷ illustrated in the example:

$$14 \div 7$$

Whole number division problems may be presented as a single-digit whole number divided evenly by a single-digit whole number (e.g., 8 ÷ 4); or perhaps by a two-digit whole number being divided evenly by a single-digit whole number (e.g., 10 ÷ 5); or in numerous other combinations of divisors and dividends.

The Division Process

The first step in reviewing the division process is to review a three-digit dividend being divided by a one-digit divisor. The divisor is a 7 and the dividend is 336.

$$7\overline{)336}$$

The first step is to see how many times that the divisor (7) can be divided into the first two numbers of the dividend (33). (If the first number in the dividend had been greater than the divisor, the first step would have been to divide it into just the first number.)

It is apparent the divisor can be divided into the first two numbers of the dividend four times. The division is made, and the 4 is placed over the second 3 in the dividend.

$$\begin{array}{r} 4 \\ 7\overline{)336} \end{array}$$

The next step is to multiply the 4 × 7 and place the result (28) below the 33 of the dividend, subtract the 28 from the 33, and bring down the result (5).

$$\begin{array}{r} 4 \\ 7\overline{)336} \\ \underline{28} \\ 5 \end{array}$$

Now bring down the next number in the numerator (6).

$$\begin{array}{r} 4 \\ 7\overline{)336} \\ \underline{28} \\ 56 \end{array}$$

The next step is to divide the divisor (7) into the 56, and place the result (8) in the quotient above the 6 in the dividend. Bring down the result (56) and place it below the 56 that was being divided.

$$\begin{array}{r} 48 \\ 7\overline{)336} \\ \underline{28} \\ 56 \\ 56 \end{array}$$

After subtracting the 56 from the 56, the problem is completed. The answer to the problem is 48.

Determining Whether a Division Problem Has Been Solved Correctly

The standard procedure for determining whether or not a division problem has been completed correctly is to multiply the answer by the divisor to see if the result is the same as the dividend. With this problem, the answer (48) should first be multiplied by the divisor (7). The procedure would appear as:

$$\begin{array}{r} 48 \\ \times\ 7 \\ \hline 336 \end{array}$$

dividend   336

Because the quotient is the same as the dividend of the problem, it can be considered that the problem has been worked correctly.

Dividing a Four-Digit Dividend by a Two-Digit Divisor

Try one more division problem. For the purpose of further review, a two-digit divisor and a four-digit dividend will be used. Following is the four-digit problem.

$$34\overline{)6188}$$

The first step is to see how many times the divisor (34) can be divided into the first two numbers of the dividend (61). It is apparent that it can go one time but not two. The division is made and the 1 is placed above the 1 of the dividend.

$$34\overline{)6188}^{\ \ 1}$$

The next step is to multiply the 1 by the 34 with the result (34) placed beneath the 61 and then subtracted from the 61.

$$\begin{array}{r} 1 \\ 34\overline{)6188} \\ 34 \\ \hline 27 \end{array}$$

The next number (first 8) is then brought down and placed onto the end of the 27, resulting in the number 278.

$$\begin{array}{r} 1 \\ 34\overline{)6188} \\ 34 \\ \hline 278 \end{array}$$

The 278 is then divided by the divisor (34). It will go eight times. The resultant 8 is placed above the 8 that was brought down.

$$\begin{array}{r} 18 \\ 34\overline{)6188} \\ 34 \\ \hline 278 \end{array}$$

The divisor 34 is then multiplied by the 8 that was placed in the quotient and the result placed below the 278.

```
        18
    ┌─────
 34)6188
     34
     ───
     278
     272
```

The next step is to subtract the 272 from the 278 and bring down the next number in the dividend (the last 8) and place it onto the end of the result from the subtraction (6).

```
        18
    ┌─────
 34)6188
     34
     ───
     278
     272
     ───
      68
```

The 68 is then divided by the divisor (34) and the result (2) is placed in the quotient. The divisor (34) is then multiplied by the 2 and the result (68) is moved down and subtracted from the dividend (68). Because there is no remainder after the subtraction is completed, the 182 in the quotient is the answer to the division problem.

```
       182
    ┌─────
 34)6188
     34
     ───
     278
     272
     ───
      68
      68
     ───
      00
```

Determining Whether the Problem Has Been Solved Correctly

The standard procedure for determining whether or not a division problem has been worked correctly will be used. With this problem, the answer (182) should first be multiplied by the divisor (34). The procedure would appear as:

```
              182
             ×34
             ────
             728
             546
Result      6,188
```

If the result is the same as the dividend of the problem, it can be considered that the problem has been worked correctly. The dividend of the problem is 6188.

Determining Whether a Dividend Can be Divided Evenly

Occasionally an individual preparing to solve a division problem wants to know whether or not the dividend can be divided evenly. Several rules can assist in determining this.

1. **Dividing evenly by 2.** A number can be divided evenly by 2 if the last digit in the number is an even number. For example, any of the following numbers can be divided evenly by 2.

    47,632   894   3,866   9,988

2. **Dividing evenly by 3.** A number can be divided by 3 if the sum of the digits of the number can be divided by 3.

    4,275   82,614   22,857

3. **Dividing evenly by 4.** If the last two digits of a number can be divided evenly by 4, or the last two digits are zeros, the entire number can be divided evenly by 4. For example, the following numbers can be divided evenly by 4.

    6,739,524   9,758,808   4,638,900

4. **Dividing evenly by 5.** If the last digit in a number is a 0 or 5, the entire number can be divided evenly by 5. For example, the following numbers can be divided evenly by 5.

    784,320   328,500   98,675

5. **Dividing evenly by 6.** A number can be divided evenly by 6 if it is an even number and the sum of the numbers can be divided evenly by 3. The following numbers can be divided evenly by 6.

    473,454   3,901,836   148,884

6. **Dividing evenly by 7.** Double the last digit of the number, then subtract the result from the remainder of the number. If the result can be divided by 7, the entire number can be divided by 7.

    Example: 36,764 $4 \times 2 = 8$

    $3676 - 8 = 3,668$ $3,668 \div 7 = 524$

    Thus, $36,764 \div 7 = 5,252$

7. **Dividing evenly by 8.** An even number can be divided evenly by 8 if the last three digits can be divided evenly by 8, or the last three digits are zeros. As an example, the following numbers can be divided evenly by 8.

    975,408   457,112   5,463,000

8. **Dividing evenly by 9.** If the sum of the digits is divisible by 9, the entire number is divisible by 9. Here are three examples.

    34,677   349,785   607,320

It's time to see whether you understood the instructions for dividing numbers. It is challenging, but solve each of the following. When you finish, check each of the problems to see whether you have worked it correctly. If it does not check out, work it again. If it still does not check out, go back and review the instructions that were given on how to work division problems.

Problem 1: Divide 135 by 5.

Problem 2: Divide 4,452 by 21.

Shortcuts for Division Problems

There are several shortcuts that can be used on division problems. They are easy to use and can be valuable at times.

DIVIDING BY 5

A number can be divided by 5 by multiplying the number by 2 and moving the decimal point one place to the left. It should be remembered that all whole numbers have an invisible decimal point at the end of the number.

Problem: Divide 443 by 5.

Solution: First multiply the dividend by 2: $443 \times 2 = 886$.
　　Next, move the decimal point one place to the left: $886 = 88.6$
　　Therefore, $443 \div 5 = 88.6$

Problem: Divide 234.7 by 5.

Solution: First, multiply the dividend by 2: $234.7 \times 2 = 469.4$
　　Next, move the decimal point one place to the left: 469.4 becomes 46.94
　　Therefore, $234.7 \div 5 = 46.94$

DIVIDING BY 10

A number can be divided by 10 by moving the decimal point one place to the left.

Problem: Divide 678.34 by 10.

Solution: Move the decimal point one place to the left: 678.34 becomes 67.834
　　Therefore, $678.34 \div 10 = 67.834$

DIVIDING BY 25

To divide a number by 25, multiply the number by 4 and move the decimal point two places to the left.

Problem: Divide 272 by 25.

Solution: First, multiply the 272 by 4. $272 \times 4 = 1088$
　　Next, move the decimal point two places to the left: 1088 becomes 10.88
　　Therefore, 272 divided by $25 = 10.88$

DIVIDING BY 50

To divide a number by 50, multiply the number by 2 and move the decimal point two places to the left.

Problem: Divide 8,143.22 by 50.

Solution: First, multiply 8,143.22 by 2: $8,143.22 \times 2 = 16,286.44$
　　Next, move the decimal point two places to the left: 16,286.44 becomes 162.8644.
　　Therefore, $8,143.22 \div 50 = 162.8644$

　　Now let's check to see whether you understood how to divide by 5, 10, 25, and 50. Work each of the following problems by the shortcut method. If you have any trouble, go back and review the procedure that explained how to do it.

Problem 1: Divide 227 by 5
Problem 2: Divide 423.72 by 5
Problem 3: Divide 582.9 by 10
Problem 4: Divide 34.78 by 25
Problem 5: Divide 567.42 by 50

Dividing Positive and Negative Integers

Follow the same procedure for the division process as previously explained. However, there is one difference: The answer may be either positive or negative.

The "sign rules" are:

1. If the dividend and the divisor have the same sign, the answer is positive.

    Examples $\quad +4\overline{)+20}^{\,+5} \quad -4\overline{)-20}^{\,+5}$

2. If the dividend and the divisor have different signs, the answer is negative.

    Examples $\quad -4\overline{)+20}^{\,-5} \quad +4\overline{)-20}^{\,-5}$

Reader's Practice Problems—4.1

1. Divide 14,661 by 27.
2. Divide 25,529 by 49.
3. Which of the following numbers can be divided evenly by 2?
    45,567   98,993   23,544
4. Which of the following numbers can be divided evenly by 3?
    42,315   89,423   17,922
5. Which of the following numbers can be divided evenly by 4?
    42,363   79,745   56,824
6. Which of the following numbers can be divided evenly by 5?
    45,795   6,240   98,734
7. Which of the following numbers can be divided evenly by 6?
    47,232   78,454   62,476
8. Which of the following numbers can be divided evenly by 8?
    98,896   45,277   26,544
9. Which of the following numbers can be divided evenly by 9?
    94,626   71,460   94,265
10. Divide 478 by 5 using the shortcut method.
11. Divide 862 by 25 using the shortcut method.
12. Divide 342 by 50 using the shortcut method.
13. If -568 is divided by $+32$, what sign will the answer have?
14. If $+7,843$ is divided by -45, what sign will the answer have?
15. If $-5,843$ is divided by -17, what sign will the answer have?

The answers are found in the Reader's Practice Problems section of the Appendix.

UNIT 1—REVIEW QUESTIONS

1. What is the term used for the number that is to be divided in a division problem?
2. What is the term used for the answer that is determined for a division problem?
3. What is the number called that is used to divide the dividend into equal parts called?
4. If a positive number is divided by a negative number, what sign will the answer be given?
5. If a negative number is divided by a negative number, what sign will the answer be given?
6. Explain how to determine whether a number can be divided evenly by 2.
7. Explain how to determine whether a number can be divided evenly by 3.
8. How can you determine whether a number can be divided evenly by 4?
9. How can you determine whether a number can be divided evenly by 5?
10. What is the best method to use to determine whether a number can be divided evenly by 6?
11. Explain how to determine whether a number can be divided evenly by 8.
12. How can you determine whether a number can be divided evenly by 9?
13. Explain the shortcut method for dividing a number by 5.
14. Explain the shortcut method for dividing a number by 25.
15. Explain the shortcut method for dividing a number by 50.

TEST 4.1

Solve the following ten division problems and prove that your answer is correct.

1. $4\overline{)72}$
2. $26\overline{)358}$
3. $77\overline{)8239}$
4. $17\overline{)2456}$
5. $11\overline{)4444}$
6. $21\overline{)3467}$
7. $+15\overline{)+45}$
8. $-15\overline{)+45}$
9. $+15\overline{)-45}$
10. $-15\overline{)-45}$

Figure 4.1

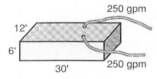

11. The swimming pool shown in Figure 4.1 holds 23,000 gallons of water. It is being filled by two hose lines, each discharging 250 gpm (gallons per minute). How long will it take to fill up the pool?

    SOLUTION HELP—Divide 23,000 by 500

12. A section of $2\frac{1}{2}$-inch hose is 50 feet in length. How many sections did it take to reach from the pumper to the fire shown in Figure 4.2?

    SOLUTION HELP—Divide 1,200 by 50

13. The nozzle being used in Figure 4.3 is discharging 195 gpm. A nozzle attached to a portable monitor being used on this fire is discharging 975 gpm. How many of the nozzles shown in Figure 4.3 will it take to discharge the same amount of water per minute as the portable monitor?

    SOLUTION HELP—Divide 975 by 195

Figure 4.2

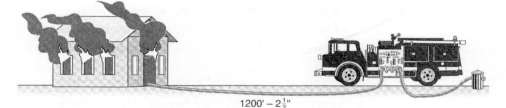

$1200' - 2\frac{1}{2}''$

Figure 4.3

350' – 1½"   195 gpm

Figure 4.4

350' – 1½"
Total friction loss 63 psi

14. A section of hose is 50 feet in length. What would be the friction loss in each of the sections of hose shown in the layout in Figure 4.4?

    SOLUTION HELP—Divide 63 by 7

15. The total friction loss in the layout shown in Figure 4.5 is 44 psi. What is the length of the hose line?

    SOLUTION HELP—Divide 44 by 8

16. A young firefighter performed 414 minutes of physical training while on duty at Station 5 during the month of June. He worked 9 shifts during the month. How many minutes of physical training did he average per shift?

    SOLUTION HELP—Divide 414 by 9

Use the following numbers to answer Questions 17 through 24.

a. 28,847	b. 76,945	c. 54,824	d. 26,823	e. 55,687
f. 47,634	g. 97,452	h. 56,843	i. 17,937	j. 72,1048

17. Which of the numbers can be divided evenly by 2? (give by the letter)
18. Which of the numbers can be divided evenly by 3?
19. Which of the numbers can be divided evenly by 4?
20. Which of the numbers can be divided evenly by 5?
21. Which of the numbers can be divided evenly by 6?
22. Which of the numbers can be divided evenly by 8?
23. Which of the numbers can be divided evenly by 9?
24. Which of the numbers cannot be divided evenly by 2, 3, 4, 5, 6, 8, or 9?
25. Using the shortcut method, divide 413 by 5.
26. Using the shortcut method, divide 234 by 25.
27. Using the shortcut method, divide 374.53 by −50.

Figure 4.5

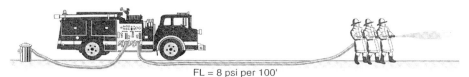

FL = 8 psi per 100'

28. $43\overline{)43.86}$

29. $54\overline{)3464.7}$

30. $+62\overline{)-73.72}$

31. $-56\overline{)-754.56}$

32. $+43\overline{)+231.35}$

33. A swimming pool holds 545,000 gallons of water. How long will it take to fill the pool if it is being filled by two sections of 3-inch hose, each discharging 600 gpm?

    SOLUTION HELP—Divide 545,000 by 1,200. Then divide the answer by 60 to find the number of hours it will take to fill the pool.

34. A section of hose measures 50 feet. Engine 3 has laid 450 feet of $2\frac{1}{2}$-inch hose reduced to 500 feet of $1\frac{1}{2}$-inch hose to reach the fire. How many sections of hose did it take to reach the fire?

    SOLUTION HELP—Add 450 to 500 and then divide the result by 50.

35. A portable monitor used at a fire is discharging 800 gpm. How many lines of $2\frac{1}{2}$-inch hose discharging 210 gpm will it take to discharge at least as much water as the portable monitor? Remember that a section of hose is 50 feet in length.

    SOLUTION HELP—Divide 800 by 210. Round your answer to the next 50 feet.

36. A paramedic assigned to Ambulance 7 worked at 24 medical emergencies during the month of June. The paramedic worked 9 shifts during the month. What was the average length of time that the paramedic worked at each emergency?

    SOLUTION HELP—Divide 24 by 9

37. The crew at Station 7 devoted 386 minutes to physical training during the month of June. The crew was on duty for 8 shifts during the month. How long did they average on each physical fitness session during the month?

    SOLUTION HELP—Divide 386 by 8

38. Use the shortcut method to divide 378 by 5.

39. Use the shortcut method to divide 744 by 25.

40. Use the shortcut method to divide 1,374 by 50.

UNIT 2—SUBTRACTION PROBLEMS

WORDS TO REMEMBER

minuend—The upper number of a subtraction problem.

subtrahend—The lower number of a subtraction problem.

difference—The answer to the problem, which is the difference between the minuend and the subtrahend.

A Review of the Subtraction Process

To solve a subtraction problem, it is necessary to determine the difference between the minuend and the subtrahend. A subtraction problem could be confused with an addition problem if the minus (−) sign is not placed to the left of the subtrahend. An example of a subtraction problem is:

$$\begin{array}{r}996 \leftarrow \text{minuend}\\-482 \leftarrow \text{subtrahend}\\\hline 514 \leftarrow \text{difference}\end{array}$$

To solve the problem, first determine the difference between the minuend and subtrahend in the ones column, then the tens column, and next the hundreds column, and so on. The following indicates the different columns:

```
h
u
n
d
r       t
e       e       o
d       n       n
s       s       e
                s

9       9       6    ← minuend
4       8       2    ← subtrahend
─────────────────
5       1       4    ← difference
```

When the digit in the minuend is smaller than the digit in the subtrahend, it is necessary to borrow a one (in reality a ten) from the column to the left and add it to the digit in the minuend of the column. In the following problem, it is necessary to borrow a one from the 9 in the hundreds column and add it to the 7 in the tens column. The first and second steps in the procedure appear as:

```
                          8 ← ← ← ← the 9 becomes an 8
borrow 1 from 9 → →       9   17   6
                      −   4    8   2
                      ─────────────
                                    4
```

The solution to the problem can then be completed by subtracting in the tens and hundreds columns.

```
        8
        9   17   6
    −   4    8   2
    ─────────────
        4    9   4
```

Proving a Subtraction Problem

The standard method for determining whether or not a subtraction problem has been completed correctly is as follows.

$$\begin{array}{r}9\ 7\ 6 \leftarrow \text{minuend}\\-\ 4\ 8\ 2 \leftarrow \text{subtrahend}\\\hline 4\ 9\ 4 \leftarrow \text{difference}\end{array}$$

Step 1: Add the answer (difference) to the number being subtracted (subtrahend).

$$494 + 482 = 976$$

Step 2: If the answer is the same as the minuend, the problem has been solved correctly. This problem has been solved correctly.

Another Example

This example demonstrates another means of using the standard method for checking a subtraction problem to see whether it has been worked correctly.

$$\begin{array}{r} 8\ 2\ 9 \\ -\ 4\ 5\ 6 \\ \hline 3\ 7\ 3 \end{array}$$

Step 1: Add the answer (difference) to the number being subtracted (subtrahend).

$$373 + 456 = 829$$

Step 2: If the answer is the same as the minuend, the problem has been solved correctly. This problem has been solved correctly.

Using the procedures just demonstrated for subtracting and proving that the subtraction problem has been solved correctly, solve all of the following and place your answer in the space provided. If the problem does not check out to be correct, rework it until it does.

Single-Digit Subtraction Problems

5	2	8	7	4	5	9	3
2	1	6	4	2	4	7	2

Double-Digit Subtraction Problems

65	34	82	45	84	42	56	72
45	23	51	34	57	26	38	17

Triple-Digit Subtraction Problems

635	839	263	727	845	562	746	835
224	317	152	416	587	374	555	444

Subtracting Negative and Positive Integers

When subtracting negative integers and positive numbers, change the sign of the number being subtracting and proceed with the problem as if it were an addition problem.

Problem: Subtract a positive number from a positive number.

Subtract +15 from +35. This can be written as:
$$\begin{array}{r} +35 \\ -\ (+15) \end{array}$$

Because the subtrahend has unlike signs, the problem can be changed to

+35 and worked as an addition problem.
$$\begin{array}{r} +35 \\ -15 \\ \hline \end{array} \qquad \begin{array}{r} +35 \\ -15 \\ \hline +20 \end{array}$$

Problem: Subtract a negative number from a positive number.

Subtract −9 from +30. This can be written as:
$$\begin{array}{r} +30 \\ -\ (-9) \end{array}$$

Change the problem to read $+30$ and proceed as if it were an addition problem
$$\begin{array}{r}+30\\+\ 9\\\hline+39\end{array}$$

Problem: Subtract a positive number from a negative number.

Subtract $+34$ from -45. This can be written as:
$$\begin{array}{r}-45\\-(+34)\\\hline\end{array}$$

Because the subtrahend has unlike signs, the problem can be changed to

-45 and worked as an addition problem.
$+34$
$$\begin{array}{r}-45\\+34\\\hline-79\end{array}$$

Problem: Subtract a negative number from a negative number.

Subtract -8 from -12. This can be written as:
$$\begin{array}{r}-12\\-(-8)\\\hline\end{array}$$

Because the subtrahend has like signs, the problem can be changed to

-12 and worked as an addition problem.
$+8$
$$\begin{array}{r}-12\\+8\\\hline-4\end{array}$$

Using the procedures just demonstrated for subtracting negative integers, solve all of the following and place your answer in the space provided.

Problem: Subtract 45 from 89. _____

Problem: Subtract -63 from $+78$. _____

Problem: Subtract $+89$ from -45. _____

Problem: Subtract -63 from -82. _____

Reader's Practice Problems—4.2

The following are all subtraction problems. Solve each of them and prove that your answer is correct.

43	79	48	88	72	645	981	645	582	645
− 27	− 45	− 37	− 45	− 27	− 87	− 95	− 53	− 58	− 99

4,567	2,365	9,821	7,925	3,345	7,845	4,532	9,054
− 342	− 678	− 789	− 153	− 2,522	− 5,798	− 2,104	− 7,009

1. Subtract $+70$ from $+90$.
2. Subtract -13 from $+52$.
3. Subtract $+22$ from -45.
4. Subtract -52 from -75.

The answers can be found in the Reader's Practice Problems section of the Appendix.

UNIT 2—REVIEW QUESTIONS

1. What is the number being subtracted in a subtraction problem called?
2. What is the number from which a number is being subtracted in a subtraction problem called?
3. What is the answer to a subtraction problem called?
4. How is a subtraction problem proven to be correct by the conventional method?
5. What sign will the answer take if a negative number is subtracted from a positive number?
6. What sign will the answer take if a positive number is subtracted from a negative number?

TEST 4.2

Solve and prove the answers to Problems 1 through 6.

1. 52 − 16
2. 2,795 − 1,546
3. 76,525 − 47,649
4. +17
   − −9
5. −320
   − +17
6. −455
   − +223

The following problems contain a number of terms common to the fire service. The meanings were explained in Chapter 2 but are repeated here to save a reader who is not familiar with fire department terminology from having to flip back through the pages to recall the terms previously used.

FL—friction loss in psi (pounds per square inch)

NP—nozzle pressure (the pressure at the nozzle tip in pounds per square inch) (psi)

BP—back pressure (the pressure in psi that is working against the pumper when the pumper is supplying water to an elevation above the pumper)

FLOW—the discharge of water from a nozzle tip or a pumper given in gallons per minute (gpm)

RPDP—the required pumper discharge pressure in psi

7. The pumper shown in Figure 4.6 is providing water to several other hose lines that have been taken off the discharge gates on the other side of the apparatus. These lines are not visible in the figure. What is the total amount of water being supplied to the other lines?

   SOLUTION HELP—Subtract 325 from 1,125

8. The elevated platform in Figure 4.7 is being supplied by two separate relay teams. The upper relay team is supplying 725 gpm to the elevated platform. How much water is being supplied to the elevated platform by the lower relay team?

   SOLUTION HELP—Subtract 725 from 1,200

Figure 4.6

Discharge 1125 gpm

Flow 325 gpm

Tip 1¼"

Figure 4.7

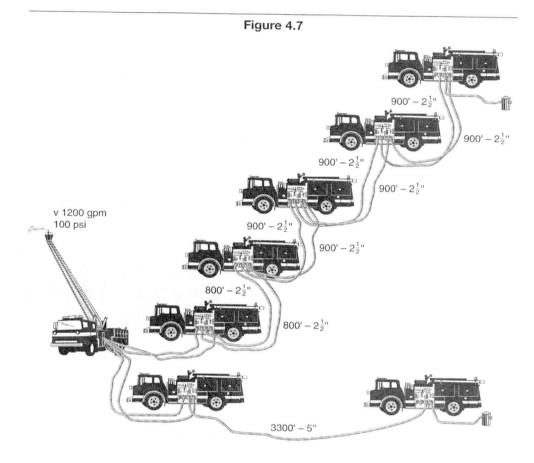

9. The pumper shown in Figure 4.8 is supplying a hose line that is laid up a hill, resulting in back pressure against the pump. The pumper has to provide sufficient pressure to provide for the friction loss in the hose, the nozzle pressure, and the back pressure. What is the back pressure in this situation?

SOLUTION HELP—Subtract 48 + 50 from 150

Figure 4.8

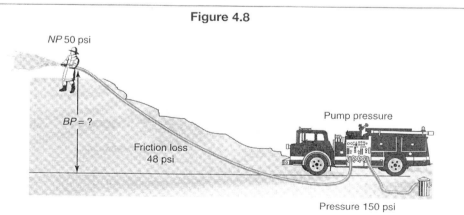

Figure 4.9

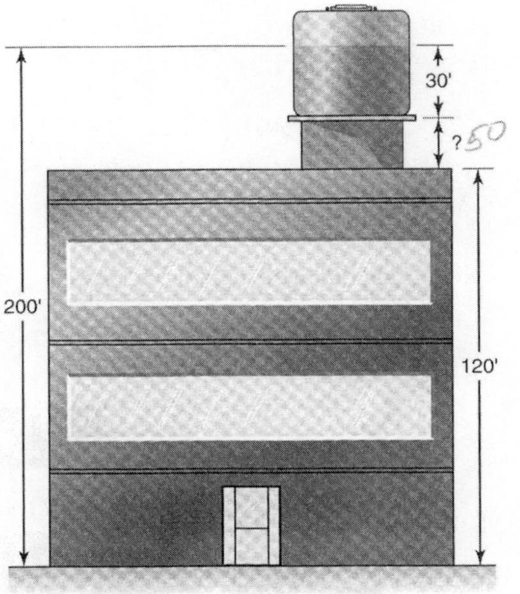

10. There is a water tank on the roof of a building that is 120 feet tall. The water tank has 30 feet of water in it. As shown in Figure 4.9, the distance from the ground to the level of the water in the tank is 200 feet. What is the distance from the bottom of the tank to the roof of the building?

    SOLUTION HELP—Subtract 120 from 170

11. The pumper shown in Figure 4.10 is discharging 650 gpm. This water is used to supply three hose lines taken off the pumper. How much water is being supplied to the lower hose line?

    SOLUTION HELP—Subtract 225 + 200 from 650

12. Two 3-inch lines are taken off the exterior standpipe to supply the portable monitor being used on the roof of an 8-story building. Each line from the pumper to the exterior standpipe inlet is 500 feet in length. According to Figure 4.11, how long are each of the 3-inch lines from the exterior standpipe outlet on the roof to the portable monitor if there are 2,100 feet of hose used in the layout?

    SOLUTION HELP—Subtract 1,500 from 2,100, then divide the answer by 2.

Figure 4.10

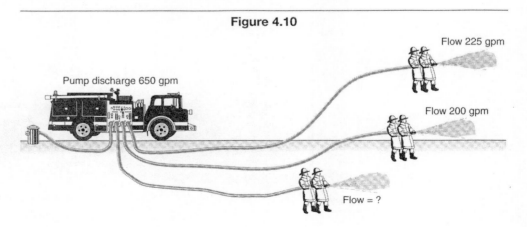

Figure 4.11

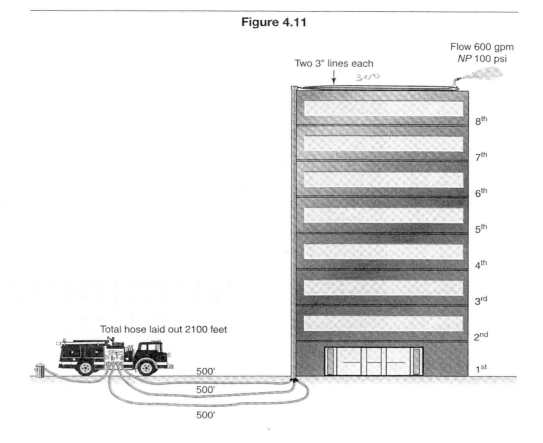

13. The pumper shown in Figure 4.12 is providing a discharge pressure to provide for the nozzle pressure, the back pressure, the friction loss in the fixed piping attached to the elevated platform, and the friction loss in the lines from the pumper to the elevated platform. What is the total friction loss in the lines from the pumper to the elevated platform?

    SOLUTION HELP—Add 80 + 26 + 20, then subtract the result from 150.

Figure 4.12

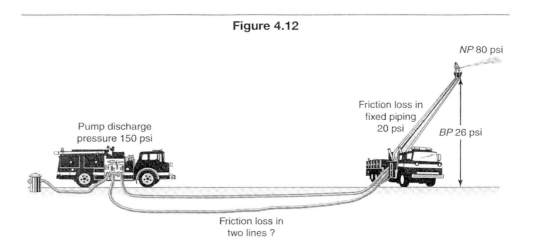

Figure 4.13

Courtesy of Rick McClure, LAFD

14. The heavy stream appliance being used on the apparatus in Figure 4.13 is discharging 895 gpm. The portable monitor on the ground is discharging 650 gpm. How many more gallons of water will the appliance on the apparatus discharge in 4 minutes than the one on the ground?

    SOLUTION HELP—Subtract 650 from 895, then multiply the answer by 4.

15. The helicopter shown in Figure 4.14 has a water tank capacity of 1,125 gallons. It drops all of the tank's water on each drop. During the day it dropped 12,375 gallons of water. Not counting the first two drops, how much water did it drop during the day?

    SOLUTION HELP—Subtract 2,250 from 12,375.

16. A line has been laid from a pumper to a fire that is located 60 feet above the pumper. The back pressure is: BP = .5H. The RPDP is 155 psi and the nozzle pressure is 50 psi. What is the friction loss?

    SOLUTION HELP—The RPDP = nozzle pressure + friction loss + back pressure Add 50 + 30 and subtract the result from 155.

17. A building that is 96 feet tall has a 30-feet water tank that is located 25 feet above the roof of the building. The tank is full of water. What is the distance from the top of the water to the ground?

    SOLUTION HELP—Add 30 + 25 + 96.

18. Three lines have been taken off a pumper. How much water is being supplied to the third line if the RPDP is 725 gpm, and 210 gallons is being supplied to the first line and 195 is being supplied to the second line?

    SOLUTION HELP—Add 210 + 195, then subtract the result from 725.

Figure 4.14

Courtesy of Rick McClure, LAFD

19. Two heavy streams are being used on a warehouse fire. One of the streams is discharging 950 gpm whereas the other stream is discharging 800 gpm. How much more water will be discharged from the heavier stream in 18 minutes than from the lighter stream?

    SOLUTION HELP—Subtract 800 from 950, then multiply the result by 18.

20. The water tank on a helicopter holds 1,200 gallons of water. During the day, it drops 19,200 gallons of water on the fire. Not counting the first three drops, how much water did it drop during the day?

    SOLUTION HELP—Subtract 3,600 from 19,200.

UNIT 3—SQUARE ROOT PROBLEMS

WORDS TO REMEMBER

radical symbol—A symbol placed over a number to indicate that the square root of the number is to be found. The symbol is: $\sqrt{}$.

exponent—A symbol to indicate how many times a number is to be multiplied by itself. It is placed to the right and raised up above the number that is to be acted upon. Examples: $2^2 = 2 \times 2 \quad 3^3 = 3 \times 3 \times 3$

A Review of the Square Root Process

To most individuals, solving a square root problem is the most difficult of all the mathematical procedures of the "so-called" basic math. It is also a cumbersome process that does not flow as easily as addition, subtraction, multiplication, and division problems. However, with a little patience and a lot of determination, you will be capable of whipping it.

In the process, you will probably also wonder why you have to learn it at all. You try to think of when you might use it in your "day-to-day" experiences, and you can't think of a single incident. However, this book is designed for use by those in the fire service. Although it is missing in the daily routine of a firefighter, square root appears many times in the standard formulas provided to firefighters for solving hydraulic problems. Therefore, here is an explanation of how the process proceeds.

The square root of a number is another number which when multiplied by itself equals the original number. For example, the square root of 16 is 4, and the square root of 81 is 9. For illustrative purposes, the problem of finding the square root of 574.274 will be used. Almost all square root problems are presented by placing the number under the square root radical symbol.

$$\sqrt{574.274}$$

The first step in the process of solving a square root problem is to place the decimal point directly above its location in the problem.

$$\sqrt{574.274}$$

The next step is to divide the problem into groups of two on both sides of the decimal point. If there is an odd number of digits to the left of the decimal point (as with this problem), then the first group in the problem will have only a single digit. If there is an odd number of digits to the right of the decimal point, as with this problem, a zero will be added to make up a group of two. In some instances it will be necessary to add additional groups of two zeros to the problem.

$$\sqrt{574.2740}$$

The first mathematical step in solving the problem is to determine a number whose square is less than or equal to the first group of two or the first number. A square means that the number is to be multiplied by itself.

In this problem, it is 2 squared ($2 \times 2 = 4$), which is less than 5. Place the 2 directly above the 5.

$$\sqrt{5\ 74.27\ 40}^{\ 2\ \ \ .}$$

Then square the 2, place the result (4) below the 5, and subtract it from the 5. Then bring down the first pair (74) and attach it to the end of the 1.

$$\begin{array}{r} 2\ \ \ . \\ \sqrt{5\ 74.27\ 40} \\ 4 \\ \hline 174 \end{array}$$

Multiply the number above the line by 2 ($2 \times 2 = 4$), add a zero to the result (40), and place it directly to the left of the 174. This is referred to as a trial divisor.

$$\begin{array}{r} 2\ \ \ . \\ \sqrt{574.2740} \\ 40\ \ \ \ \ \ \ \ 4 \\ \hline 174 \end{array}$$

Estimate how many times the trial divisor (40) will go into the 174. It appears that it will go 4 times ($4 \times 40 = 160$) but not 5 times ($5 \times 40 = 200$). Place the 4 on the top line to the right of the 2 and over the 74. Then replace the 0 in the trial divisor with a 4.

$$\begin{array}{r} 2\ \ 4. \\ \sqrt{5\ 74.27\ 40} \\ 4 \\ \hline \end{array}$$
$$44 \quad 174$$

The next step is to multiply the trial divisor (now the divisor) by the 4 in the answer space, place the result below the 174, and subtract. However, it is apparent when the multiplication is completed, that the result (176) is greater than the 174. This makes it necessary to go back and use a 3 in place of the 4. *The purpose of using the 4 when it was too large is to indicate to you that it is easy to make a mistake when working a square root problem and what to do to correct it. Correcting it is not difficult.*

$$\begin{array}{r} 2\ \ 3. \\ \sqrt{5\ 74.27\ 40} \\ 4 \\ \hline \end{array}$$
$$43 \quad 174$$

Now multiply the divisor (43) by the 3 and place the result below the 174 and subtract.

$$\begin{array}{r} 2\ \ 3. \\ \sqrt{5\ 74.27\ 40} \\ 4 \\ \hline 174 \\ 129 \\ \hline 45 \end{array}$$
$$43$$

From this point on, the process is a repeat of what was done in the above step. Bring down the next pair of numbers (27) and place them at the end of the 45 (4527). Multiply the numbers above the problem (23) by 2, place the result (46) to the left of the 4527, and add a zero on the end (460). This is a new trial divisor.

$$\begin{array}{r} 2\ \ 3. \\ \sqrt{5\ 74.27\ 40} \\ 4 \\ \hline 174 \\ 129 \\ \hline 4527 \end{array}$$
$$43$$
$$460$$

Estimate how many times the 4527 can be divided by the 460. It appears that it will go 9 times. Place the 9 above the 27 in the problem, change the 0 in the trial divisor to a 9, and multiply the 469 by 9. Place the result (4221) below the 4527 and subtract.

$$\begin{array}{r} 2\ \ 3.\ \ 9 \\ \sqrt{5\ 74.27\ 40} \\ 4 \\ \hline 174 \\ 129 \\ \hline 4527 \\ 4221 \\ \hline 306 \end{array}$$
$$43$$
$$469$$

Repeat the process. Bring down the next pair (40) and place it to the right of the 306. Multiply the numbers in the answer (239) by 2 and place the result (478) to the left of the 30640. Then add a zero to the 478. This is the new trial divisor.

$$
\begin{array}{r}
23.9 \\
\sqrt{574.2740} \\
4 \\
43\overline{174} \\
129 \\
469\overline{4527} \\
4221 \\
4780\overline{30640}
\end{array}
$$

Estimate how many times the 30640 can be divided by the 4780. It appears that it will go 6 times. Place the 6 in the answer above the 40, and replace the 0 in the trial divisor with a 6. Multiply the divisor (4786) by the 6 that was placed in the answer, place the result (28716) below the 30640, and subtract.

$$
\begin{array}{r}
23.96 \\
\sqrt{574.2740} \\
434 \\
\overline{174} \\
469129 \\
\overline{4527} \\
47864221 \\
\overline{30640} \\
28716 \\
\overline{1924} \text{ (this is the remainder)}
\end{array}
$$

Solving a Two-Digit Square Root Problem

Most of the hydraulic formulas used in the fire service that include determining the square root of a number are associated with finding the square root of the nozzle pressure. The most standard nozzle pressure for handheld straight tips is 50 psi. Consequently, an example of solving a two-digit number is to determine the square root of 50. The problem would appear as follows:

$$\sqrt{50}$$

The first step involves the placement of the decimal point and the dividing of the problem into groups of two on both sides of the decimal point.

This problem does not have a decimal point; however, all whole numbers have a hidden decimal point to the right of the last digit. With the whole number 50, there is a hidden decimal point to the right of the zero. Make it visible by placing it there. Now add two groups of zeros to the right of the located decimal point.

$$\sqrt{50.\,00\,00}$$

Determine the largest number that when squared (multiplied by itself) will produce a result less than 50. It should be apparent that the number is 7. Place the 7 above the 50, square it, and place the result (49) below the 50. Then subtract it from the 50. The result will be 1. Bring down the first pair of zeros and place them to the right of the 1.

$$\begin{array}{r} 7. \\ \sqrt{50.\,00\,00} \\ 49 \\ \hline 1\,00 \end{array}$$

Multiply the 7 above the problem by 2, place the result (14) to the left of the 100, and add a 0. This is the trial divisor.

$$\begin{array}{r} 7. \\ \sqrt{50.\,00\,00} \\ 49 \\ 140 1\,00 \end{array}$$

The next step would be to divide the trial divisor into the 100; however, it is obvious that it is larger than the 100. Because the division cannot be performed, place a 0 above the first pair of zeros and bring down the next pair.

$$\begin{array}{r} 7.\,0 \\ \sqrt{50.\,00\,00} \\ 49 \\ 140 1\,00\,00 \end{array}$$

Now multiply the 70 by 2, place the result (140) to the left of the 10000, and add a zero. This is your trial divisor.

$$\begin{array}{r} 7.\,0 \\ \sqrt{50.\,00\,00} \\ 49 \\ 1400 1\,00\,00 \end{array}$$

Estimate how many times the 10000 can be divided by the 1400. It is 7. Place the 7 above the last pair of zeros, and change the last zero of 1400 to a 7 (result 1407). Now multiply the 1407 by the last 7 in the answer and place the result (9849) below the 10000 and subtract.

$$\begin{array}{r} 7.\,0\,7 \\ \sqrt{50.\,00\,00} \\ 49 \\ 1407 1\,00\,00 \\ 98\,49 \\ \hline 51 \end{array}$$

From the process for determining the square root of 50, the answer is 7.07. If you are a member of the fire service or planning to be a member of a fire department, remember this answer. It will be used frequently when working hydraulic problems for handheld lines. The formulas where it is used most often to determine the square root of the nozzle pressure is:

$$\text{Discharge} = 29.7 \ \ D^2 \sqrt{P}$$

Proving a Square Root Problem

To demonstrate the procedure for proving whether or not a solved square root problem has been worked correctly, the first square root problem worked will be used. Following is the problem in its completed form.

```
                    2  3. 9  6
                 √5 74.27 40
                   4
         43       174
                  129
        469       4527
                  4221
       4786      30640
                 28716
                  1924  (remainder)
```

The traditional method of proving a square root problem is to square the answer, then add the remainder. The result should be the same as the original problem. However, because of the rounding process, sometimes the answer is close but not exact. The process would appear as follows:

```
                           23.96
                        ×  23.96
                          14376
                          21564
                           7188
                           4792
                         5740816
  add the remainder    +    1924
                         574.2740
```

Because the result is the same as the original problem, the answer is correct. However, it can be seen that multiplying 23.96 × 23.96 is not a simple multiplication problem. It is one that very few people would enjoy tackling.

Painful as it may be, use the illustrations of how to work a square root problem and prove that it has been worked correctly. Solve the following problems and place your answer in the space provided. If your problem does not check out as being worked correctly, rework it until it does. However, when you square the number you receive on working the problem, the answer may be close to the original problem but not exact. This can result from rounding off the number.

Problem: Determine the square root of 45.

$$\sqrt{45} = _____$$

Problem: Determine the square root of 80.

$$\sqrt{80} = _____$$

If you are a firefighter, or preparing to become one, and cannot work the square root problems, there is one more word of advice. It is time for you to memorize one more chart. It will not be any more difficult than memorizing the multiplication charts. Here it is. You will find it quite valuable during your firefighting career.

Nozzle Pressure	Square Root
40	6.32
45	6.71
50	7.07
55	7.42
60	7.75
70	8.37
80	8.94
90	9.49

Reader's Practice Problems—4.3

Solve the following two square root problems and prove each of them by the conventional method.

$$\sqrt{647.8542} \qquad \sqrt{70}$$

The answers are found in the Reader's Practice Problems section of the Appendix.

UNIT 3—REVIEW QUESTIONS

1. What does a radical symbol placed over a number indicate?
2. What does an exponent placed over a number indicate?
3. What is the first step in solving a square root problem?
4. What is the second step in solving a square root problem?
5. How is a square root problem proved to be worked correctly by the conventional method?

TEST 4.3

Solve the following problems and prove that your answers are correct by using the conventional method. If you had trouble working the square root problems in Unit 3, this test may prove to be very difficult. Do your best and be sure to learn to work square root problems prior to taking an apparatus test for a fire department.

1. $\sqrt{6279}$
2. $\sqrt{60}$
3. Determine the square root of 3.742.
4. What is the square root of 70?
5. What is the square root of 345.64?
6. Determine the square root of the nozzle pressure shown in Figure 4.15.

Figure 4.15

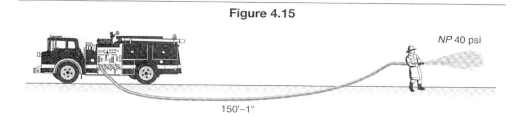

Figure 4.16

1½" Tip
NP 80 psi

7. The horizontal reach of a solid stream from a nozzle tip can be determined by the formula:

$$S = \sqrt{(HF)(P)}$$

where S = horizontal reach
HF = horizontal factor
P = nozzle pressure

The horizontal factor for a 1½-inch tip is 104.

Problem: Determine the horizontal reach of the stream being discharged from the ladder pipe shown in Figure 4.16.

SOLUTION HELP—First, multiply the horizontal factor × the nozzle pressure. Next, determine the square root of the answer.

8. *Problem:* The large tank shown in Figure 4.17 has a hole in its side from which water is being discharged. The hole in the side is located 62 feet below the surface of the water. The 62 feet is referred to as the head. The speed with which the water is flowing out the hole is referred to as the velocity flow. The formula for determining the velocity flow is:

$$V = 8\sqrt{H}$$

where V = velocity flow
H = head

What is the velocity flow from the hole in the side of the tank shown in Figure 4.17?

SOLUTION HELP—The head is 62. Determine the square root of the head. Next, multiply 8 × the answer.

9. *Problem:* The formula for determining the velocity flow from a nozzle tip is:

$$V = 12.14\sqrt{P}$$

V = velocity flow
P = nozzle pressure

Figure 4.17

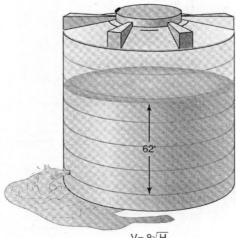

$V = 8\sqrt{H}$

SOLVING WHOLE NUMBER DIVISION, SUBTRACTION, AND SQUARE ROOT PROBLEMS 85

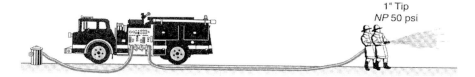

Figure 4.18

What is the velocity flow from the nozzle tip being used in Figure 4.18?

SOLUTION HELP—First, determine the nozzle pressure (see the chart earlier in this chapter). Next, multiply 12.14 by the square root of 50.

10. **Problem:** The formula for determining the distance of a vertical stream being discharged from a nozzle tip is:

$$S = \sqrt{(VF)(P)}$$

where S = vertical stream
       VF = vertical factor
       P = nozzle pressure

The vertical factor for the $\frac{3}{4}$-inch tip being used in Figure 4.19 is 95. What is the vertical reach of the stream?

SOLUTION HELP—First, multiply the vertical factor by the nozzle pressure (see the chart in this chapter). Next, determine the square root of the answer.

11. **Problem:** The nozzle pressure on the nozzle shown in Figure 4.20 is 80 psi. What is the square root of the nozzle pressure?

SOLUTION HELP—See the chart in this chapter.

12. Occasionally, in the fire service, it is desired to replace two or more tips with a single tip that will most nearly provide the same amount of water as the two or

Figure 4.19

Figure 4.20

Courtesy of Rick McClure, LAFD

more tips. The formula used to determine this is referred to as the equivalent nozzle pressure formula. The formula is:

$$END = \frac{\sqrt{D1^2 + D2^2 + D3^2 + \text{etc.}}}{8}$$

where   END = equivalent nozzle diameter
D1 = diameter of 1st tip in eighths
D2 = diameter of 2nd tip in eighths
D3 = diameter of 3rd tip in eighths

Problem: What single-size tip will most nearly discharge the same amount of water as the combined discharge from a $\frac{7}{8}$-inch tip, a $1\frac{1}{8}$-inch tip, and a $1\frac{1}{4}$-inch tip?

(***Note:*** Try to work the problem. However, because it is a complicated problem, the solution is shown below.)

$$END = \frac{\sqrt{D1^2 + D2^2 + D3^2 + \text{etc.}}}{8}$$

where   D1 = $\frac{7}{8}$-inch tip − in eighths = 7
D2 = $1\frac{1}{8}$-inch tip − in eighths = 9
D3 = $1\frac{1}{4}$-inch tip − in eighths = 10

then

$$END = \frac{\sqrt{7^2 + 9^2 + 10^2}}{8}$$

$$= \frac{\sqrt{49 + 81 + 100}}{8}$$

$$= \frac{\sqrt{230}}{8}$$

$$= \frac{15.17}{8}$$

= It most nearly equals a $1\frac{7}{8}$-inch tip; however, the $1\frac{7}{8}$-inch tip would not discharge quite as much water.

Chapter Summary

This chapter was divided into three units: Unit 1 was devoted to the division of whole numbers, Unit 2 the subtraction of whole numbers, and Unit 3 working square root problems. Each unit provided the definition of words used in the process, a review of the procedure for solving the problem, and an illustration of how to prove the answer to a problem by the conventional method. Each unit also included both standard and word problems for the reader to solve as well as review questions.

CHAPTER 5
WORKING WITH FRACTIONS

KEY TERMS

canceling
common fractions
composite number
cross-multiplication
denominator
difference

equivalent fraction
fractions
improper fraction
like fractions
minuend
minuend fraction

mixed number
numerator
proper fraction
subtrahend
subtrahend fraction
unlike fractions

OBJECTIVES AND GOALS

Upon completing this chapter, the reader should be able to:

- Define the words fractions, numerator, denominator, equivalent fractions, improper fractions, and mixed numbers.
- Solve addition problems involving fractions.
- Solve multiplication problems involving fractions.
- Solve subtraction problems involving fractions.
- Solve division problems involving fractions.
- Determine the common denominator of an addition or subtraction problem.
- Reduce a fraction to its lowest term.

UNIT 1

INTRODUCTION

Working with fractions involves learning the language associated with fractions and the methods of adding, subtracting, multiplying, and dividing fractions. This introductory unit should provide the reader with a good basic foundation for working with the various facets of mathematical operations involving fractions. The first step is to acquaint the reader with the words used.

WORDS TO REMEMBER

fractions—Used to represent different parts of a whole number or object.
numerator—The top number in a fraction.
denominator—The bottom number in a fraction.
equivalent fraction—A fraction that represents the same amount as another fraction although the two fractions have different numbers.
improper fraction—A fraction whose numerator is greater than its denominator.
mixed number—A whole number and a fraction with the fraction having a value of less than 1.

Introduction to Fractions

Fractions are part of a whole. A whole, such as the number 1 or perhaps a physical object such as a pie, can be divided into a number of parts. Each of the parts will be identified as a fraction. Regardless of how many parts, or the size of the parts, that a whole is divided into, when all of the fractions represented by the parts are totaled, they will always equal the whole.

Note, however, that the parts of a whole may also be illustrated by decimals or by percentages. A comparison of the three is shown below.

Fraction	Decimal	Percent
1/8	.125	12.5%
1/4	.25	25%
3/8	.375	37.5%
1/2	.5	50%
5/8	.625	62.5%
3/4	.75	75%
7/8	.875	87.5%

Take a look at a pie that is divided up by fractions. The pie shown in Figure 5.1 is divided up several different ways. However, note that the fractions representing the various parts in each of the divisions, when added, equal the whole pie.

Fractions are presented as a top number over a bottom number, divided by a line called the fraction bar. This, in essence, presents the top number being divided by the bottom number. The top number is referred to as the numerator whereas the bottom number is referred to as the denominator. In fraction form, it would appear as:

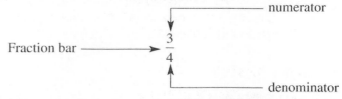

Fractions can be added, multiplied, subtracted, or divided. The process of accomplishing each of these is presented in the remainder of this chapter.

Types of Common Fractions

Common fractions are identified as *proper fractions, improper fractions, like fractions, unlike fractions,* and *mixed numbers.*

A ***proper fraction*** has a numerator smaller than the denominator. Examples of proper fractions are: $\frac{3}{4}, \frac{2}{9}, \frac{3}{7}, \frac{4}{11}$, and so on.

Figure 5.1

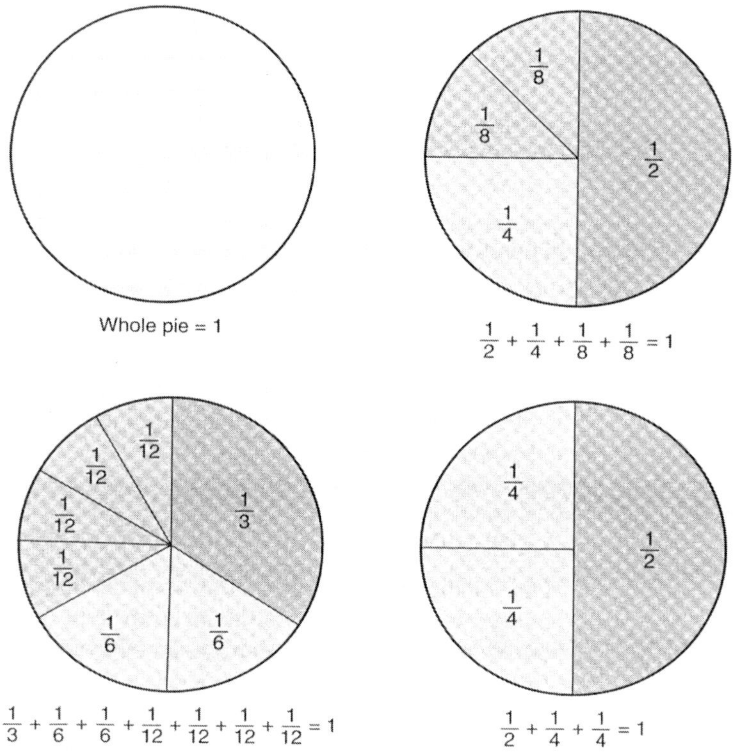
Whole pie = 1
1/2
1/4
1/8
1/8
1/2 + 1/4 + 1/8 + 1/8 = 1
1/3
1/6
1/6
1/12
1/12
1/12
1/12
1/3 + 1/6 + 1/6 + 1/12 + 1/12 + 1/12 + 1/12 = 1
1/2
1/4
1/4
1/2 + 1/4 + 1/4 = 1

An ***improper fraction*** has a numerator equal to or larger than the denominator. An example of each is: $\frac{2}{2}$ and $\frac{5}{3}$.

Like fractions are two or more fractions that have the same denominator. Examples of like fractions are $\frac{2}{7}$ and $\frac{5}{7}$.

Unlike fractions are two or more fractions that have different denominators. Examples of unlike fractions are $\frac{3}{7}$ and $\frac{3}{11}$.

Mixed numbers are a whole number or numbers together with a fraction. Examples of mixed numbers are $3\frac{1}{4}$ and $23\frac{1}{2}$.

Related Principles of Whole Numbers

When working with fractions, it is important to be familiar with a couple of concepts concerning whole numbers.

1. Any number that has two or more digits is either a prime number or a product of two or more prime numbers. If a number is a product of two or more prime numbers, it is called a ***composite number***. Some examples of composite numbers are 15 (3×5), 32 ($2 \times 2 \times 2 \times 2 \times 2$), and 165 ($3 \times 5 \times 11$). This principle will be used when the process for finding a common denominator is examined.

   To determine the prime numbers in a composite number, start with the 2, then the 3, then the 5, and so on up the scale.

Problem: Determine the prime numbers in the composite number 420.

Solution: First try the 2. Because 420 ends in an even number, it can be divided by 2.

$$420 \div 2 = 210.$$

Second, because 210 is an even number, another 2 can be removed.

$$210 \div 2 = 105$$

105 is not an even number. Therefore, all of the 2's have been removed. Now try a 3.

$$105 \div 3 = 35$$

Because 35 cannot be divided by 3, try 5.

$$35 \div 5 = 7$$

Therefore, the prime numbers in 420 are $2 \times 2 \times 3 \times 5 \times 7$.

2. The value of a number is not changed when it is multiplied by 1. For example, $55 \times 1 = 55$, and $743 \times 1 = 743$. This also applies to fractions. The value of a fraction is not changed when it is multiplied by 1. This principle will be used when changing a fraction to an equivalent fraction.
3. The number 1 can be changed to an improper fraction without changing its value. For example, $1 = \frac{3}{3}$, and $1 = \frac{7}{7}$.

Changing a Fraction to an Equivalent Fraction

One of the standard operations associated with adding and subtracting fractions is to convert a common fraction to an equivalent fraction. As an example, if it becomes necessary to change the fraction $\frac{3}{4}$ to an equivalent fraction having a denominator of 12, the following procedure would be used.

$$\frac{3}{4} = \frac{}{12}$$

Divide the denominator of the fraction to be changed into the denominator of the equivalent fraction: $12 \div 4 = 3$. Multiply the result by the numerator of the fraction to be changed: $3 \times 3 = 9$. Make the result of the multiplication the numerator of the equivalent fraction. The result would appear as:

$$\frac{3}{4} = \frac{9}{12}$$

Changing a Fraction to a Higher or Lower Term

Some mathematical processes require a fraction to be changed to a higher or lower term. Changing a fraction's term refers to changing its form without changing its value. Changing a fraction's term to a higher term can be achieved by multiplying both the numerator and the denominator by the same number. In essence, this is the same as multiplying the fraction by 1. For example, $\frac{2}{3}$ can be changed to a higher term by multiplying both the numerator (2) and the denominator (3) by 2. The result will appear as:

$$\frac{2 \times 2}{3 \times 2} = \frac{4}{6}$$

Changing a fraction to a lower term can be achieved by dividing both the numerator and denominator by the same number. For example, $\frac{3}{12}$ can be changed to a lower term by dividing both the numerator and the denominator by 3. The process would appear as:

$$\frac{3 \div 3}{12 \div 3} = \frac{1}{4}$$

A simple check can be made to determine if two fractions are of equal value. The process is called cross-multiplication. **Cross-multiplication** refers to a system where the numerator on the left side of an equal sign is multiplied by the denominator on the right side, and the numerator on the right side is multiplied by the denominator on the left side. The process would appear as:

$$\frac{3}{12} = \frac{1}{4} = \frac{3 \times 4}{1 \times 12} = \frac{12}{12}$$

If the results are the same, the two fractions are of equal value.

Converting a Mixed Number to an Improper Fraction

Converting a mixed number to an improper fraction is a relatively simple task. The procedure is to multiply the whole number by the denominator of the fraction, add the numerator of the fraction, and place the result over the original denominator. The procedure is illustrated in the following example.

Problem: Change the mixed number $5\frac{1}{3}$ to an improper fraction.

Solution: $5\frac{1}{3}$ Multiply $3 \times 5 = 15$

Add the 1: $15 + 1 = 16$

Place the 16 over the 3

$$\frac{16}{3}$$

Converting an Improper Fraction to a Mixed Number

The procedure is to divide the numerator of the improper fraction by the denominator of the fraction. The quotient will be a whole number. Place the remainder over the original denominator of the improper fraction, then place the result on the end of the obtained whole number. The result is the mixed number.

Problem: Convert $\frac{19}{4}$ to a mixed number.

Solution: $\frac{19}{4}$ Divide the 19 by the 4. The result is a 4 with a remainder of 3.

Place the 3 over the 4 and add the result to the 4 obtained by the previous division.

Result: $4\frac{3}{4}$

Converting a Decimal to a Fraction

To convert a decimal to a fraction, place the decimal over 1, and then repeatedly multiply both the numerator and the denominator of the formed fraction by 10 until the numerator is a whole number.

Problem: Convert .25 to a fraction.

Solution: $\dfrac{.25}{1} = \dfrac{(.25)(10)}{(1)(10)} = \dfrac{2.5}{10} = \dfrac{(2.5)(10)}{(10)(10)} = \dfrac{25}{100}$

Reduce $\frac{25}{100}$ to its lowest term by dividing both the numerator and denominator by 25

$$\frac{25 \div 25}{100 \div 25} = \frac{1}{4}$$

Converting Minutes to a Fraction

Occasionally it is desirable to convert minutes to a fraction. Minutes can be converted to a fraction by using the following steps:

1. Place the number of minutes over 60. This will result in a fraction having a denominator of 60.
2. If the minutes end in zero, cancel the zero in both the numerator and the denominator. If it does not end in a zero, proceed to Step 3.
3. Reduce the fraction to its lowest term.

Problem: Convert 20 minutes into a fraction.

Solution:

1. Place the 20 over 60. This produces the fraction $\frac{20}{60}$.
2. Because both the numerator and denominator end in zero, cancel the zeros.

$$\frac{20}{60} = \frac{2}{6}$$

3. Reduce $\frac{2}{6}$ to its lowest term by dividing both the 2 and the 6 by 2.

$$\frac{2 \div 2}{6 \div 2} = \frac{1}{3}$$

The answer indicates that 20 minutes is equivalent to $\frac{1}{3}$ of an hour.

Problem: Convert 25 minutes into a fraction.

Solution:

1. Place the 25 over the 60. This produces the fraction $\frac{25}{60}$. Proceed to Step 3.
2. Reduce the fraction $\frac{25}{60}$ to its lowest term by dividing both the numerator (25) and the denominator (60) by 5.

$$\frac{25 \div 5}{60 \div 5} = \frac{5}{12}$$

The answer indicates that 25 minutes is equivalent to $\frac{5}{12}$ of an hour.

Following are some problems that are provided to see whether you understand the material you have been exposed to so far in this chapter. Work each of the problems and place your answer in the space provided. If you cannot work any of the problems, go back and review the material that was recently presented.

Problem 1. Change $\frac{2}{3}$ to an equivalent fraction that has a denominator of 9.

$$\frac{2}{3} = \frac{}{9} \qquad \underline{}$$

Problem 2. Change the fraction $\frac{7}{8}$ to a higher fraction that has a denominator of 24.

$$\frac{7}{8} = \frac{}{24} \qquad \underline{}$$

Problem 3. Change $\frac{12}{16}$ to a lower fraction that has a denominator of 4.

$$\frac{12}{16} = \frac{}{4}$$ _____

Problem 4. Convert the mixed number $8\frac{1}{4}$ to an improper fraction.

$$8\frac{1}{4} = \frac{}{4}$$ _____

Problem 5. Convert $\frac{13}{4}$ to a mixed number.

$$\frac{13}{4} = \frac{}{4}$$ _____

Problem 6. Convert the decimal .75 to a fraction and reduce the fraction to its lowest terms.

$$.75 = $$ _____

Problem 7. Convert 36 minutes to a fraction.

36 minutes equals _____

Reader's Practice Problems—5.1

1. Change the fraction $\frac{2}{3}$ to an equivalent fraction with a denominator of 18.
2. Change the fraction $\frac{3}{4}$ to a higher fraction having a denominator of 8.
3. Change the fraction $\frac{9}{12}$ to a lower fraction having a denominator of 4.
4. Convert the mixed number $7\frac{1}{2}$ to an improper fraction.
5. Convert the improper fraction $\frac{16}{3}$ to a mixed number.
6. Convert the decimal .42 to a fraction and reduce the fraction to its lowest terms.
7. Convert 50 minutes to a fractional portion of an hour.
8. Convert 36 minutes to a fractional portion of an hour.
9. Provide the numerator of the denominator 5 so that the developed fraction is a proper fraction.
10. Break the following composite numbers into their prime numbers.
    8, 18, 30, 105, 108, 165

The answers are in the Reader's Practice Problems section of the Appendix.

UNIT 1—REVIEW QUESTIONS

1. What is the definition of a fraction?
2. A fraction consists of one number over another number. What is the top number called?
3. What is the bottom number of a fraction called?
4. What is an equivalent fraction?
5. What is the definition of an improper fraction?
6. What does the term "mixed number" refer to?

7. What is a composite number?
8. How do you convert a mixed number to an improper fraction?
9. How do you convert an improper fraction to a mixed number?
10. How do you change a fraction to an equivalent fraction?
11. Name the different types of common fractions.
12. How do you change a decimal to a fraction?
13. What is the definition of a like fraction?
14. What is the definition of a proper fraction?
15. What is the definition of an unlike fraction?

TEST 5.1

1. Convert $\frac{17}{3}$ to a mixed number.
2. Convert $\frac{9}{2}$ to a mixed number.
3. Convert $6\frac{2}{3}$ to an improper fraction.
4. Convert $7\frac{3}{5}$ to an improper fraction.
5. Change $\frac{3}{11}$ to an equivalent fraction with a denominator of 22.
6. Change $\frac{5}{7}$ to an equivalent fraction with a denominator of 21.
7. What prime numbers make up the composite number 42?
8. What prime numbers make up the composite number 195?
9. What is the next larger fraction that is a like fraction to $\frac{3}{5}$?
10. What is the next larger fraction that is a like fraction to $\frac{7}{9}$?
11. What number is missing from the following?

$$\frac{2}{5} = \frac{}{15}$$

12. What number is missing from the following?

$$\frac{1}{3} = \frac{}{12}$$

13. What number is missing from the following?

$$\frac{2}{7} = \frac{}{21}$$

14. Are $\frac{3}{7}$ and $\frac{3}{5}$ like fractions?
15. Are $\frac{2}{3}$ and $\frac{18}{29}$ equal fractions?
16. Convert 40 minutes into a fraction.
17. Convert 50 minutes into a fraction.
18. Convert $\frac{19}{3}$ to a mixed number.
19. Convert $6\frac{7}{8}$ to an improper fraction.
20. Change $\frac{7}{21}$ to an equivalent fraction with a denominator of 33.
21. What prime numbers make up the number 66?
22. What prime numbers make up the number 455?

UNIT 2—ADDITION OF FRACTIONS

Introduction

The first step that has to be completed in order to add two or more fractions is to ensure that each of the fractions has the same denominator. This is referred to as finding a common denominator.

Finding a common denominator is generally not difficult if the denominators of all the fractions to be added have a single digit. If the addition involves only two fractions, the simplest method of finding the common denominator is to multiply the denominators of the two fractions. For example, if the problem is to add $\frac{1}{2} + \frac{1}{3}$, multiply $2 \times 3 = 6$. Then change each of the fractions to be added to fractions having an equal value and a denominator of 6. The operation would appear as:

$$\frac{1}{6} = \underline{}$$ divide the 6 by the 2, multiply the resultant 3 by the numerator of the $\frac{1}{2}$ (1), and place the result over the denominator (6)

$$\frac{1}{2} = \frac{3}{6}$$

Repeat the process for $\frac{1}{3}$. The result would appear as:

$$\frac{1}{3} = \frac{2}{6}$$

Now the two equivalent fractions can be added. The process is to add the numerators and place the result over the common denominator (6).

$$\frac{3}{6} + \frac{2}{6} = \frac{3+2}{6} = \frac{5}{6}$$

The next step after determining the sum of the two fractions is to reduce the fraction obtained to its lowest term. When attempting to do this, there is an important principle that should be remembered. The principle is: *If the numerator and denominator do not have a common prime factor, the fraction has been reduced to its lowest term.* The most useful prime numbers to be used for checking are 2, 3, 5, 7, and 11. It might be useful to review the information in Chapter 1 with reference to prime numbers.

The 5 in the numerator of the answer $\frac{5}{6}$ is a prime number and the same prime number is not in the 6. Therefore, it can be concluded that the fraction $\frac{5}{6}$ has been reduced to its lowest term.

Another point to remember: If a fraction is a proper fraction and the denominator is a prime number, the fraction has been reduced to its lowest terms. For example, 11 is a prime number. Therefore, if 11 is the denominator of a fraction, any digit below 11 can be the numerator and the fraction will be reduced to its lowest term. For example, the fractions $\frac{1}{11}, \frac{2}{11}, \frac{3}{11}, \frac{4}{11}, \frac{5}{11}, \frac{6}{11}, \frac{7}{11}, \frac{8}{11}, \frac{9}{11}$, and $\frac{10}{11}$ are all reduced to their lowest term.

This principle is even more valuable when the denominator is a larger number. For example, 91 is a prime number. Therefore, if 91 is the denominator of a proper fraction, any number lower than 91 can be the numerator and the fraction will be reduced to its lowest term. It would not be necessary to check to see whether the numerator can be divided by a prime number.

The process of determining the lowest common denominator is a little bit more difficult when three fractions are added. Of course, it is always possible to determine a common denominator by multiplying the three denominators. This would produce a common denominator but not likely a denominator that would result in the fraction obtained by the addition process being reduced to its lowest term. It is also likely that the common denominator might be very large. Following is an example of how this would work.

Problem: Add $\dfrac{2}{9} + \dfrac{1}{3} + \dfrac{4}{15}$

Solution: Multiply the denominators to obtain a common denominator. $9 \times 3 \times 15 = 405$.

Now set up the problem by changing each of the fractions to be added to equivalent fractions having a common denominator of 405.

$$\dfrac{2}{9} = \dfrac{90}{405}$$

$$\dfrac{1}{3} = \dfrac{135}{405}$$

$$\dfrac{4}{15} = \dfrac{108}{405}$$

Next, add the ~~denominators~~ *numerators* and place the result over the 405.

$$\dfrac{90 + 135 + 108}{405} = \dfrac{333}{405}$$

Reduce $\dfrac{333}{405}$ to its lowest term.

First, divide the numerator and denominator by 3.

$$\dfrac{333 \div 3}{405 \div 3} = \dfrac{111}{135}$$

Now divide the numerator and denominator of $\dfrac{111}{135}$ by 3.

$$\dfrac{111 \div 3}{135 \div 3} = \dfrac{37}{45}$$

Because 37 is a prime number, the fraction has been reduced to its lowest term.

Following is another method of determining a common denominator that is also the lowest common denominator of the fractions being added. The method makes use of the prime numbers. The same problem will be used to demonstrate the process.

Add $\dfrac{2}{9} + \dfrac{1}{3} + \dfrac{4}{15}$

The first step is to change each of the denominators to a composite of their prime numbers. The change would appear as follows:

$$\dfrac{2}{9} = \dfrac{(2)}{(3)(3)}$$

$$\dfrac{1}{3} = \dfrac{(1)}{(3)}$$

$$\dfrac{4}{15} = \dfrac{(2)(2)}{(3)(5)}$$

The next step is to add to each of the denominators any prime numbers found in the denominators of either of the other two fractions that is not found in its denominator. For example, the denominator of the fraction $\dfrac{2}{9}$ contains two 3's but does not contain a 5 that is found in the fraction $\frac{4}{15}$. The denominator of the fraction $\frac{1}{3}$ is missing one of the 3's that is found in the $\frac{2}{9}$ fraction and the 5 that is found in the $\frac{4}{15}$ fraction. The fraction $\frac{4}{15}$ is missing one of the 3's found in the fraction $\frac{2}{9}$. The missing prime numbers should be added to each of the denominators. When a prime number is added to a denominator, it should also be added to the numerator. The result of adding it to both the numerator and the denominator is the same as multiplying the fraction affected by 1. As previously discussed, multiplying a whole number or a fraction by 1 does not change the value of the number or fraction multiplied. The completed operation would appear as:

$$\frac{2}{9} = \frac{(2)}{(3)(3)} = \frac{(2)(5)}{(3)(3)(5)} \quad \text{A 5 has been added.}$$

$$\frac{1}{3} = \frac{(1)}{(3)} = \frac{(1)(3)(5)}{(3)(3)(5)} \quad \text{A 3 and a 5 have been added.}$$

$$\frac{4}{15} = \frac{(2)(2)}{(3)(5)} = \frac{(2)(2)(3)}{(3)(5)(3)} \quad \text{A 3 has been added.}$$

The next step is to complete the multiplication of both the numerators and the denominators and place the result of the multiplication of the numerators over the result of the multiplication of the denominators. The result would appear as follows:

$$\frac{2}{9} = \frac{(2)}{(3)(3)} = \frac{(2)(5)}{(3)(3)(5)} = \frac{10}{45}$$

$$\frac{1}{3} = \frac{(1)}{(3)} = \frac{(1)(3)(5)}{(3)(3)(5)} = \frac{15}{45}$$

$$\frac{4}{15} = \frac{(4)}{(3)(5)} = \frac{(2)(2)(3)}{(3)(3)(5)} = \frac{12}{45}$$

Then add the numerators and place the result over the denominator.

$$\frac{10 + 15 + 12}{45} = \frac{37}{45}$$

Because the 37 in the numerator is a prime number, the fraction that is the sum of the addition problem is reduced to its lowest term.

If neither the numerator nor the denominator of the completed fraction is a prime number, then it is advantageous to use the canceling method to achieve the objective. For example, suppose that the answer had been $\frac{36}{45}$ rather than $\frac{37}{45}$. The procedure then is to change both the numerator and the denominator into their respective prime numbers and cancel. The procedure would appear as:

$$\frac{36}{45} = \frac{(2)(2)(3)(3)}{(3)(3)(5)} \quad \text{now cancel} = \frac{(2)(2)(3)(3)}{(3)(3)(5)} = \frac{4}{5}$$

It should be noted that once the prime numbers common to both the numerator and the denominator have been removed (in this problem the two 3's) the remaining fraction $\left(\frac{4}{5}\right)$ has been reduced to its lowest term.

The situation is a little different when both of the denominators have two digits. For example, add $\frac{9}{42}$ and $\frac{21}{30}$. The easiest method of adding these two fractions is to first reduce the two fractions to their lowest terms. However, for the purpose of demonstrating another method of solving the addition, the traditional method of multiplying the two denominators to establish a common denominator is used.

If the two denominators are multiplied, the result will be 1,260 (42 × 30 = 1,260). This would be the common denominator for the two fractions. To solve the problem, it would be set up as:

$$\frac{9}{42} = \frac{270}{1,260}$$

$$\frac{21}{30} = \frac{882}{1,260}$$

Now add the numerators (270 + 882) and place them over the common denominator (1,260). The process would appear as:

$$\frac{270 + 882}{1,260} = \frac{1,152}{1,260}$$

The next step would be to reduce the $\frac{1,152}{1,260}$ to its lowest term. The process to use would be:

Using the divisibility rules, it can be seen that both the numerator and denominator can be divided by both 2 and 3. These two prime numbers can be combined and 6 can be used as the first divisor.

$$\frac{1,152 \div 6}{1,260 \div 6} = \frac{192}{210}$$

Using the divisibility rules again, it can be seen that both the numerator and denominator can again be divided by 6. The result would appear as follows.

192 ÷ 6 = 32   Not only has the addition problem been solved, but 210 ÷ 6 = 35, the resultant fraction, has been reduced to its lowest term because there is not a prime number remaining that is common to both the numerator and denominator.

Addition of Mixed Numbers

The addition of some mixed numbers is a simple process of adding the whole numbers and adding the fractions separately.

Problem: Add $5\frac{3}{7} + 2\frac{2}{7}$

Solution: First add the whole numbers: 5 + 2 = 7

Then add the fractions: $\frac{3}{7} + \frac{2}{7} = \frac{5}{7}$

Final step. Combine the whole number result to the fraction result.

$$7 + \frac{5}{7} = 7\frac{5}{7}$$

For other mixed number addition problems, the problem can be solved by converting one of the fractions to an equivalent fraction having a denominator the same as the other fraction, and then proceeding the same as with the previous problem.

Problem: Add $6\frac{1}{4} + 7\frac{1}{2}$

Solution: First change the fraction $\frac{1}{2}$ to the equivalent fraction $\frac{2}{4}$. Then proceed as before.

$$6\frac{1}{4}$$

$$7\frac{2}{4}$$

First, add the whole numbers: $6 + 7 = 13$

Second, add the fractions: $\frac{1}{4} + \frac{2}{4} = \frac{3}{4}$

Then, combine the results: $13\frac{3}{4}$

Some problems involving the addition of mixed numbers are more complex. One procedure to use is:

1. Convert each of the mixed numbers to an improper fraction.
2. Add the improper fractions. The sum will be an improper fraction.
3. Convert the improper fraction to a mixed number.
4. Reduce the fraction portion of the mixed number to its lowest term.

Problem: Add $2\frac{2}{3} + 1\frac{3}{5}$

Solution:

1. Convert each of the mixed numbers to an improper fraction

$$2\frac{2}{3} = \frac{8}{3} \quad 1\frac{3}{5} = \frac{8}{5}$$

2. Convert the improper fractions to fractions having a common denominator.

$$\frac{8}{3} \text{ Change to: } \frac{40}{15}$$

$$\frac{8}{5} \text{ Change to: } \frac{24}{15}$$

3. Add the numerators: $40 + 24 = 64$. Place the result over the common denominator: $\frac{64}{15}$

4. Convert the improper fraction to a mixed number.

$$\frac{64}{15} = 4\frac{4}{15}$$

5. Normally, the fraction would then be reduced to its lowest terms. However, with this problem, it is already at its lowest term because the 4 and the 15 do not have a common prime number.

Problem: Add $4\frac{1}{8} + 3\frac{3}{5} + 2\frac{1}{2}$

Solution:

1. Convert each of the mixed numbers to an improper fraction.

$$4\frac{1}{8} = \frac{33}{8} \quad 3\frac{3}{5} = \frac{18}{5} \quad 2\frac{1}{2} = \frac{5}{2}$$

2. Convert the improper fractions to fractions having a common denominator.

$$\frac{33}{8} \text{ Change to: } \frac{165}{40}$$

$$\frac{18}{5} = \frac{144}{40}$$

$$\frac{5}{2} = \frac{100}{40}$$

3. Add the numerators. $165 + 144 + 100 = 409$

4. Place the result over the common denominator.

$$\frac{409}{40}$$

5. Convert the improper fraction to a mixed number.

$$\frac{409}{40} = 10\frac{9}{40}$$

6. Reduce the fraction portion of the mixed number to its lowest term. Because there is no prime number common to both the numerator and the denominator, the fraction is reduced to its lowest term.

It is time to see if you understand how to multiply fractions and mixed numbers. Work each of the following problems, and reduce your answer so that the fraction is reduced to its lowest term. If you have any trouble, go back to the examples and review how the problem should have been worked.

Problem 1. Add $\frac{3}{10}$ and $\frac{2}{5}$.
Problem 2. Add $\frac{7}{5}$ and $\frac{2}{3}$.
Problem 3. Add $\frac{4}{5} + \frac{1}{3} + \frac{7}{30}$
Problem 4. Add $\frac{5}{9} + \frac{3}{7} + \frac{11}{63} + \frac{2}{7}$
Problem 5. Add $3\frac{5}{7} + 7\frac{3}{7}$
Problem 6. Add $9\frac{2}{3} + 6\frac{7}{17}$

Reader's Practice Problems—5.2

Solve all of the following addition problems. Reduce each answer to its lowers terms.

$$\frac{2}{3} + \frac{3}{5} \qquad \frac{1}{2} + \frac{3}{4} \qquad \frac{3}{11} + \frac{1}{3} \qquad \frac{3}{4} + \frac{1}{2} + \frac{1}{3} \qquad \frac{11}{12} + \frac{3}{8} + \frac{1}{4}$$

$$\frac{15}{16} + \frac{11}{12} + \frac{7}{8} \qquad \frac{1}{2} + \frac{1}{3} + \frac{3}{5} + \frac{6}{7} \qquad \frac{11}{16} + \frac{5}{6} + \frac{2}{7} + \frac{1}{3}$$

$$2\frac{3}{4} + 5\frac{6}{7} + 9\frac{1}{2} \qquad 6\frac{1}{8} + 4\frac{2}{3} + 8\frac{5}{12} \qquad 17\frac{7}{8} + 12\frac{1}{2} + 9\frac{7}{16}$$

The answers are found in the Reader's Practice Problems section of the Appendix.

UNIT 2—REVIEW QUESTIONS

1. What is the first step that should be taken when adding two or more fractions?
2. What is the easiest way to convert two fractions so that they both have a common denominator?
3. When can you be sure that a fraction has been reduced to its lowest terms?
4. What is the simplest method of adding some mixed numbers?
5. What would be the first step when adding $5\frac{2}{3} + 3\frac{1}{7} + 6\frac{3}{5}$?
6. What are the steps to be taken when adding complex mixed numbers?
7. What should be remembered about a fraction whose denominator is a prime number?

TEST 5.2

1. Determine the lowest common denominator for: 3, 4, and 9.
2. Determine the lowest common denominator for: 10, 15, and 21.
3. Convert the improper fraction $\frac{11}{7}$ to a mixed number.
4. Convert the mixed number $4\frac{2}{3}$ to an improper fraction.
5. Convert the improper fraction $\frac{35}{4}$ to a mixed number.
6. Convert the mixed number $3\frac{2}{3}$ to an improper fraction.

Determine the sum of each of the following. Express the answers in their lowest terms.

7. $\frac{2}{3} + \frac{3}{5}$
8. $\frac{9}{13} + \frac{2}{9}$
9. $\frac{5}{21} + \frac{2}{15}$
10. $\frac{1}{2} + \frac{3}{4} + \frac{5}{6}$
11. $4\frac{1}{8} + 3\frac{1}{3}$
12. $1\frac{1}{2} + 2\frac{2}{3} + 3\frac{2}{5}$
13. $4\frac{1}{9} + 7\frac{4}{9}$

Note to the reader: Many of the word problems in this test and the other tests in this chapter are not true-life situations. The reason for this is that the problems are given using fractions whereas similar true-life problems would be given in whole numbers. The change has been made to provide the reader with the experience of working with fraction word problems.

Figure 5.2

Work schedule	
Monday	$8\frac{1}{2}$ hours
Tuesday	8 hours
Wednesday	$8\frac{3}{4}$ hours
Thursday	9 hours
Friday	$8\frac{1}{2}$ hours
Saturday	4 hours

14. A paramedic ambulance unit runs out of Station 5. Figure 5.2 lists one paramedic's working hours for the first week of March. The labor contract requires that every paramedic receive time and a half for all hours over 8 in a single day and all hours over 40 in a single week. How many overtime hours did the paramedic work during the week?

    SOLUTION HELP—Total $\frac{1}{2}$ hour on Monday + $\frac{3}{4}$ on Wednesday + 1 on Thursday + $\frac{1}{2}$ on Friday + 4 on Saturday.

15. On a busy Friday in May, Engine 5 responded to five emergencies. On the first response they ran $2\frac{1}{3}$ miles, on the second $3\frac{1}{2}$ miles, on the third only $\frac{7}{8}$ of a mile, on the fourth $5\frac{1}{2}$ miles, and on the fifth $4\frac{1}{3}$ miles. What was the total distance they traveled on the five alarms?

    SOLUTION HELP—Add $2\frac{1}{3} + 3\frac{1}{2} + \frac{7}{8} + 5\frac{1}{2} + 4\frac{1}{3}$

16. The members of the Friendship Fire Department were required to perform physical exercises every day they were on duty. On the first four shifts of June, the "A" shift spent $\frac{3}{4}$ hour, $\frac{5}{6}$ hour, $\frac{2}{3}$ hour, and $\frac{2}{3}$ hour doing the required exercises. How many total hours did they devote to exercise training during these four shifts?

    SOLUTION HELP—Add $\frac{3}{4} + \frac{5}{6} + \frac{2}{3} + \frac{2}{3}$

17. To keep in physical shape, firefighter Jane Thomas liked to walk on her days off. On three consecutive days she walked $4\frac{1}{2}$ miles, $3\frac{2}{3}$ miles and $3\frac{3}{4}$ miles. What was the total number of miles she walked in the three days?

    SOLUTION HELP—Add $4\frac{1}{2} + 3\frac{2}{3} + 3\frac{3}{4}$

18. In a single day, Truck Company 3 responded to five fire emergencies. They were on the first call for 50 minutes, the second for 45 minutes, the third for 30 minutes, 40 minutes on the fourth, and for 50 minutes on the fifth call. Convert the time spent on each of the calls to fractions and add the fractions to determine the total time spent on the five calls.

    SOLUTION HELP—First, convert 50 minutes, 45 minutes, 30 minutes, 40 minutes, and 50 minutes to fractions. Then add the conversions.

    Add $\frac{5}{6} + \frac{3}{4} + \frac{1}{2} + \frac{2}{3} + \frac{5}{6}$

19. The Peacetown Fire Department had three working fires in a one-month period. On the first fire, they were on the scene for 3 hours and 25 minutes. They spent 2 hours and 45 minutes on the second fire, and 4 hours and 10 minutes on the third. How much time did they spend on the three fires? Express your answer in a mixed fraction number.

    SOLUTION HELP—Add $3\frac{5}{12}$ (3 hours 25 minutes) $+ 2\frac{3}{4}$ (2 hours 45 minutes) $+ 4\frac{1}{6}$ (4 hours 10 minutes)

20. Engine Company 5 received a fire call early on Tuesday morning. It took the company $2\frac{3}{4}$ minutes to travel to the fire location, $3\frac{1}{3}$ minutes to lay out the hose lines, $5\frac{1}{3}$ minutes to knock down the fire, and $25\frac{1}{2}$ minutes to complete the overhaul operations. The return trip to quarters took $5\frac{1}{4}$ minutes. How much time (expressed in a mixed fraction) did the company devote to this fire from the receipt of the alarm to the return to quarters?

    SOLUTION HELP—Add $2\frac{3}{4} + 3\frac{1}{3} + 5\frac{1}{3} + 25\frac{1}{2} + 5\frac{1}{4}$

21. Engine 3 carries 1,200 feet of $2\frac{1}{2}$-inch hose, 500 feet of 2-inch hose, and 300 feet of $1\frac{1}{2}$-inch hose. At a small structure fire they used 400 feet of their $2\frac{1}{2}$-inch hose, 250 feet of their 2-inch hose, and 200 feet of their $1\frac{1}{2}$-inch hose. What fraction of the total amount of hose they carried was used at this fire?

    SOLUTION HELP—First, determine how much hose they carried by adding 1,200 + 500 + 300. Next, determine how much they used on the fire by adding 400 + 250 + 200. Then place the amount they used on the fire (850) over the amount they carried (2,000) to form the fraction $\frac{850}{2,000}$. Reduce this fraction to its lowest term.

    $$\frac{850}{2,000} = \frac{17}{40}$$

22. Three engine companies worked on a fire at 726 South Main Street. At the height of the fire, Engine 2 discharged $320\frac{1}{2}$ gpm of water, Engine 7 discharged $157\frac{1}{8}$ gpm, and Engine 11 discharged $185\frac{1}{3}$ gpm. What was the total amount of water (in gpm) discharged by the three companies at the height of the fire?

    SOLUTION HELP—Add $320\frac{1}{2} + 157\frac{1}{8} + 185\frac{1}{3}$ (reduce answer to lowest term)

23. The formula to use to determine the Required Pump Discharge Pressure for the layout shown in Figure 5.3 is:

    $$RPDP = FL + NP$$

    where   FL = friction loss in the hose line
    NP = nozzle pressure

    What is the RPDP for this hose layout?

    SOLUTION HELP—From Figure 5.3, add $103\frac{1}{3}$ (FL) $+ 49\frac{1}{2}$ (NP)

Figure 5.3

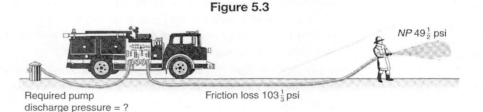

Required pump discharge pressure = ?

Friction loss $103\frac{1}{3}$ psi

NP $49\frac{1}{2}$ psi

Figure 5.4

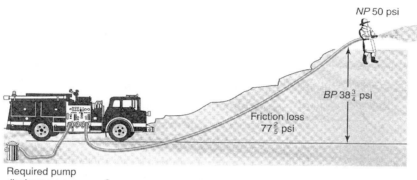

24. The formula to use to determine the Required Pump Discharge Pressure for the layout shown in Figure 5.4 is:

$$RPDP = FL + NP + BP$$

where FL = friction loss in the hose line
 NP = nozzle pressure
 BP = back pressure

What is the RPDP for this hose layout?

SOLUTION HELP—From Figure 5.4, add $77\frac{2}{5}$ (FL) + $38\frac{3}{4}$ (BP) + 50 (NP)

25. $\dfrac{3}{4} + \dfrac{7}{8}$

26. $\dfrac{3}{10} + \dfrac{5}{6}$

27. $\dfrac{7}{10} + \dfrac{2}{3} + \dfrac{1}{2}$

28. $1\dfrac{1}{2} + 3\dfrac{1}{3} + \dfrac{5}{6}$

29. The "A" shift on Engine 4 devoted three days to making fire prevention inspections during the month of June. On the first day they devoted 3 hours and 20 minutes on this effort. The second day they devoted 2 hours and 45 minutes, and on the third day they devoted $3\frac{1}{2}$ hours. Expressed in fractions, how many hours did they spend on fire prevention inspections?

SOLUTION HELP—First, change the minutes to fractions. On the first day, change 3 hours 20 minutes to $3\frac{1}{3}$ hours; on the second day, change 2 hours 45 minutes to $2\frac{3}{4}$ hours; and on the third day, use the $3\frac{1}{2}$ hours. Now add $3\frac{1}{3} + 2\frac{3}{4} + 3\frac{1}{2}$.

30. Firefighter Williams devoted a portion of each of his days off to walking. During the month of October, he walked $2\frac{1}{3}$ hours the first day, $1\frac{1}{2}$ hours the second day, and $3\frac{1}{2}$ hours the third day. How many hours did he walk during the month of October?

SOLUTION HELP—Add $2\frac{1}{3} + 1\frac{1}{2} + 3\frac{1}{2}$

UNIT 3—SUBTRACTION OF FRACTIONS

WORDS TO REMEMBER

The words used in the subtraction of fractions were introduced in a previous chapter. However, to save the reader the inconvenience of having to search back through previous chapters to find the words, they are repeated here.

minuend—The upper number of a subtraction problem.

subtrahend—The lower number of a subtraction problem.

difference—The answer to the problem, which is the difference between the minuend and the subtrahend.

In this section, the fraction to be subtracted is referred to as the ***subtrahend fraction*** and the fraction from which the subtrahend fraction is to be subtracted is called the ***minuend fraction***.

Subtraction of Common Fractions

The procedure for subtracting common fractions is very similar to the procedure for adding common fractions. The first step that has to be completed to is to make sure that each of the fractions has a common denominator. If the denominators are not the same, convert the fractions to equivalent fractions whose denominators are the same. The next step is to subtract the numerators and place the difference over the denominator.

Following are a couple of problems where the denominators are the same.

Problem: Subtract $\frac{3}{7}$ from $\frac{5}{7}$

Solution: $\frac{5}{7} - \frac{3}{7} = \frac{5-3}{7} = \frac{2}{7}$

Problem: Subtract $\frac{4}{11}$ from $\frac{7}{11}$

Solution: $\frac{7}{11} - \frac{4}{11} = \frac{7-4}{11} = \frac{3}{11}$

An example of a problem where the denominators are different:

Problem: Subtract $\frac{3}{4}$ from $\frac{7}{8}$

Solution: Convert the $\frac{3}{4}$ to an equivalent fraction having a denominator of 8.

$$\frac{3}{4} = \frac{6}{8}$$

Set up and solve the problem.

$$\frac{7}{8} - \frac{6}{8} = \frac{7-6}{8} = \frac{1}{8}$$

Subtracting Mixed Numbers

Subtracting mixed numbers is not much different from adding mixed numbers. If both of the mixed numbers have fractions with a common denominator, the procedure is to subtract the whole numbers and the fractions individually to obtain the desired result. For example, subtract $2\frac{2}{7}$ from $5\frac{3}{7}$.

Problem: $5\frac{3}{7} - 2\frac{2}{7}$

Solution: First, subtract the whole numbers: $5 - 2 = 3$
Second, subtract the fractions: $\frac{3}{7} - \frac{2}{7} = \frac{1}{7}$
Combine the results: $3 + \frac{1}{7} = 3\frac{1}{7}$

With other mixed numbers subtraction problems, the procedure is to convert one of the fractions to an equivalent fraction having the same denominator as the other fraction. For example, subtract $6\frac{1}{4}$ from $7\frac{1}{2}$.

Problem: $7\frac{1}{2} - 6\frac{1}{4}$

Solution: First, convert the fraction $\frac{1}{2}$ to the equivalent fraction $\frac{2}{4}$. Then proceed as with the previous problem. The problem becomes:

$$7\frac{2}{4} - 6\frac{1}{4}$$

First, subtract the whole numbers: $7 - 6 = 1$
Next, subtract the fractions: $\frac{2}{4} - \frac{1}{4} = \frac{1}{4}$
Combine the results: $1 + \frac{1}{4} = 1\frac{1}{4}$

As with the addition of mixed numbers, some complex mixed numbers subtraction problems require a more structured procedure. The following problem will be used to explain the procedure.

Problem: Subtract $3\frac{1}{2}$ from $8\frac{6}{7}$

1. The first step is to convert each of the fractions to equivalent fractions having a common denominator.

$$\frac{6}{7} = \frac{12}{14}$$

$$\frac{1}{2} = \frac{7}{14}$$

2. If the numerator of the minuend fraction is larger than the numerator of the subtrahend fraction, subtract the subtrahend numerator from the minuend numerator, place the answer over the common denominator, and reduce the fraction to its lowest term.

$$\frac{12}{14} - \frac{7}{14} \quad \frac{12-7}{14} = \frac{5}{14}$$

Because there is no prime number common to both the numerator and denominator, the fraction has been reduced to its lowest term.

3. Subtract the subtrahend whole number from the minuend whole number and add the result to the fraction obtained in Step 2.

$$8 - 3 = 5$$

Combine the results: $5 + \frac{5}{14} = 5\frac{5}{14}$

Problem: Subtract $5\frac{1}{2}$ from $13\frac{2}{7}$

1. The first step is to convert each of the fractions to equivalent fractions having a common denominator.

$$\frac{2}{7} = \frac{4}{14}$$

$$\frac{1}{2} = \frac{7}{14}$$

2. If the numerator of the minuend fraction is smaller than the numerator of the subtrahend fraction, remove a 1 from the minuend whole number, change it to a fraction having the same denominator as the minuend fraction, and add the two.

$$13 - 1 = 12 \quad 1 = \frac{14}{14} \quad \frac{14 + 4}{14} = \frac{18}{14}$$

3. Subtract the numerator of the subtrahend fraction from the numerator of the equivalent fraction obtained from the 1, and place the result over the common denominator. Reduce the fraction obtained to its lowest term.

$$\frac{18}{14} - \frac{7}{14} \quad \frac{18 - 7}{14} = \frac{11}{14}$$

Because the 11 is a prime number, the fraction has been reduced to its lowest term.

4. Subtract the subtrahend whole number from the reduced minuend whole number and combine the result with the fraction.

$$12 - 5 = 7$$

Combine the results: $7 + \frac{11}{14} = 7\frac{11}{14}$

The next section of this unit is the Reader's Practice Problems. However, prior to turning you loose on these problems, it is probably wise to see if you have understood the instructions on subtracting fractions. Work the following problems. If you have any trouble working any of them correctly, go back and review the instructions given. Be sure to reduce your answer to its lowest term.

Problem 1. Subtract $\frac{5}{13}$ from $\frac{11}{13}$

Problem 2. Subtract $\frac{5}{6}$ from $\frac{11}{12}$

Problem 3. Subtract $3\frac{1}{3}$ from $7\frac{2}{3}$

Problem 4. Subtract $7\frac{1}{5}$ from $9\frac{1}{3}$

Reader's Practice Problems—5.3

Subtract $\frac{4}{9}$ from $\frac{8}{9}$ Subtract $\frac{9}{12}$ from $\frac{5}{6}$ Subtract $\frac{2}{3}$ from $\frac{6}{7}$

Subtract $2\frac{1}{3}$ from $7\frac{2}{3}$ Subtract $6\frac{2}{9}$ from $12\frac{11}{12}$ Subtract $5\frac{9}{11}$ from $13\frac{2}{11}$

Subtract $65\frac{7}{8}$ from $110\frac{1}{3}$

Subtract each of the following.

$$\frac{17}{19} \qquad \frac{11}{14} \qquad \frac{4}{7} \qquad 4\frac{3}{4} \qquad 15\frac{9}{10} \qquad 8\frac{1}{3}$$
$$-\frac{11}{19} \qquad -\frac{11}{28} \qquad -\frac{2}{5} \qquad 2\frac{1}{4} \qquad 5\frac{2}{5} \qquad -6\frac{7}{9}$$

The answers are found in the Reader's Practice Problems section of the Appendix.

Figure 5.5

UNIT 3—REVIEW QUESTIONS

1. What is the upper number of a subtraction problem involving common fractions called?
2. What is the lower number of a subtraction problem involving common fractions called?
3. What is the answer to a subtraction problem involving common fractions called?
4. With what other procedure involving common fractions is the subtraction of common fractions similar?
5. What is the general procedure used for subtracting mixed numbers?

TEST 5.3

Reduce all fraction answers to their lowest term.

1. Subtract $\frac{7}{11}$ from $\frac{9}{11}$.
2. Subtract $\frac{9}{17}$ from $\frac{16}{17}$.
3. Subtract $\frac{1}{3}$ from $\frac{9}{11}$.
4. Subtract $\frac{1}{4}$ from $\frac{7}{9}$.
5. Subtract $5\frac{1}{4}$ from $8\frac{3}{4}$.
6. Subtract $4\frac{4}{5}$ from $15\frac{2}{9}$.
7. Subtract $\frac{13}{17}$ from $\frac{16}{17}$.
8. Subtract $\frac{1}{3}$ from $\frac{3}{4}$.
9. Subtract $3\frac{1}{2}$ from $7\frac{7}{8}$.
10. Subtract $4\frac{8}{9}$ from $7\frac{2}{3}$.
11. A young firefighter has established a goal of running $5\frac{1}{2}$ miles each day. On this date he has run $3\frac{1}{6}$ miles (Figure 5.5). How many more does he have to run to reach his goal?

    SOLUTION HELP—Subtract $3\frac{1}{6}$ from $5\frac{1}{2}$

12. How much more water does tank A in Figure 5.6 hold than tank B?

    SOLUTION HELP—Subtract $96\frac{3}{4}$ from $112\frac{1}{3}$

Figure 5.6

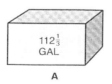

Figure 5.7

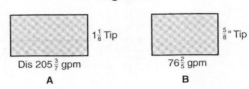

13. On three consecutive days, a paramedic on the ambulance that ran out of Station 20 worked a total of $26\frac{1}{3}$ hours. On the first day she worked $8\frac{1}{3}$ hours. On the second day she worked $8\frac{1}{2}$ hours. How long did she work on the third day?
    SOLUTION HELP—Add $8\frac{1}{3}$ (first day) + $8\frac{1}{2}$ (second day), then subtract the total of the two days from the total for the three days ($26\frac{1}{3}$).

14. How many more gpm of water is being discharged from nozzle tip A in Figure 5.7 than from nozzle tip B?
    SOLUTION HELP—Subtract $76\frac{2}{5}$ gpm (tip B) from $205\frac{3}{7}$ gpm (tip A)

15. A fire company is required to devote $28\frac{1}{3}$ hours a month making fire prevention inspections. So far this month they have completed $21\frac{2}{7}$ hours of inspection work. How many more hours are they required to make inspections?
    SOLUTION HELP—Subtract $21\frac{2}{7}$ (completed) from $28\frac{1}{3}$ (required)

16. It is estimated that the fire in a large storage tank containing gasoline would burn for $52\frac{1}{2}$ hours. It has been $39\frac{1}{6}$ hours since the fire began. How much longer is it estimated that the fire will burn?
    SOLUTION HELP—Subtract $39\frac{1}{6}$ (burned) from $52\frac{1}{2}$ (estimated to burn)

17. A young firefighter is enrolled in an education program that includes $23\frac{1}{12}$ hours of hands-on training. She has completed $18\frac{2}{3}$ hours of the required training. How many more hours is she required to do?
    SOLUTION HELP—Subtract $18\frac{2}{3}$ (completed) from $23\frac{1}{12}$ (total hours)

18. What is the distance from point D to point E in Figure 5.8?
    SOLUTION HELP—Subtract $80\frac{2}{3}$ (point D) from $121\frac{1}{2}$ (point E)

19. What is the distance from point B to point D in Figure 5.8?
    SOLUTION HELP—Subtract $40\frac{2}{7}$ (point B) from $80\frac{2}{3}$ (point D)

20. What is the distance from B to point E in Figure 5.8?
    SOLUTION HELP—Subtract $40\frac{2}{7}$ (point B) from $121\frac{1}{2}$ (point E)

Reduce all fractions to their lowest terms. All problems are subtraction problems.

21. $\dfrac{7}{8}$
    $\dfrac{3}{4}$

22. $\dfrac{5}{7}$
    $\dfrac{2}{7}$

23. $3\dfrac{3}{7}$
    $1\dfrac{7}{14}$

24. $5\dfrac{5}{11}$
    $3\dfrac{4}{11}$

25. $1\dfrac{11}{12}$
    $1\dfrac{3}{4}$

Figure 5.8

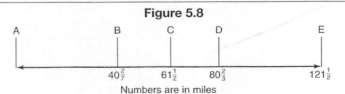

Numbers are in miles

26. A paramedic established a goal of running 3 miles on each of her days off. Today she ran $2\frac{7}{8}$ miles. How short was she of reaching her goal?

    SOLUTION HELP—Subtract $2\frac{7}{8}$ from 3

27. An elevated tank has $4,978\frac{5}{6}$ gallons of water in it. When full, it has 5,000 gallons. How much more water does it need to be full?

    SOLUTION HELP—Subtract $4,978\frac{5}{6}$ from 5,000

28. A pumper is supplying water to two $2\frac{1}{2}$-inch lines. One of the lines is discharging $309\frac{5}{6}$ gallons and the other is discharging $298\frac{2}{3}$ gallons. How much more is the first line discharging?

    SOLUTION HELP—Subtract $298\frac{2}{3}$ from $309\frac{5}{6}$

29. A newly appointed firefighter has enrolled in a course that requires 40 hours of hands-on training. So far he has completed $29\frac{5}{6}$ hours. How much more time does he have to complete?

    SOLUTION HELP—Subtract $29\frac{5}{6}$ from 40

30. A young firefighter has decided to run a distance of $4\frac{1}{2}$ miles on each of his off-duty days. So far today he has run $3\frac{7}{8}$ miles. How much more does he have to run to reach his goal?

    SOLUTION HELP—Subtract $3\frac{7}{8}$ from $4\frac{1}{2}$

UNIT 4—MULTIPLICATION OF FRACTIONS

Multiplication of Common Fractions

The multiplication of common fractions is a simple process of multiplying the numerators and multiplying the denominators. The problems generally involve small or relatively small numbers. Many of the multiplications can be completed mentally without reverting to the use of pencil and paper.

Two concepts should be remembered when multiplying two common fractions:

1. The fraction obtained by the multiplication of two fractions is generally smaller than either of the two fractions. This concept is opposite of what is obtained when two whole numbers are multiplied. For example, $\frac{3}{4}$ multiplied by $\frac{1}{2} = \frac{3}{8}$. The $\frac{3}{8}$ is smaller than either the $\frac{3}{4}$ or the $\frac{1}{2}$. However, when 3 is multiplied by 5, the result is 15, which is larger than either the 3 or the 5.

2. The × symbol used to indicate a multiplication can be replaced by the word "of". This is illustrated in Figure 5.9. The circle to the left has been cut in half. The multiplication problem is to multiply $\frac{1}{2} \times \frac{1}{2}$, which equals $\frac{1}{4}$. Using the term $\frac{1}{2}$ of $\frac{1}{2}$, it can be seen in the right-hand circle that the $\frac{1}{2}$ has been cut in half, which illustrates that $\frac{1}{2}$ of $\frac{1}{2} = \frac{1}{4}$.

Figure 5.9

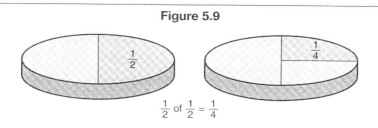

The Multiplication Process

The multiplication of common fractions is different than the adding and subtracting of common fractions. With the addition and subtraction processes, it is necessary to determine a common denominator. This is not necessary with the multiplication process. The procedure is to multiply the numerators and denominators separately without making any changes in the fractions. This is true whether two fractions are being multiplied or more than two fractions are being multiplied. However, it is still necessary to reduce the fraction obtained to its lowest term. Following are several examples.

Problem: Multiply $\dfrac{1}{2} \times \dfrac{2}{3}$

Solution: $\dfrac{1}{2} \times \dfrac{2}{3} = \dfrac{1 \times 2}{2 \times 3} = \dfrac{2}{6} = \dfrac{1}{3}$

Problem: Multiply $\dfrac{3}{4} \times \dfrac{2}{3} \times \dfrac{5}{7}$

Solution: $\dfrac{3}{4} \times \dfrac{2}{3} \times \dfrac{5}{7} = \dfrac{3 \times 2 \times 5}{4 \times 3 \times 7} = \dfrac{30}{84} = \dfrac{5}{14}$

Problem: Multiply $\dfrac{4}{7} \times \dfrac{1}{2} \times \dfrac{1}{4} \times \dfrac{2}{3}$

Solution: $\dfrac{4}{7} \times \dfrac{1}{2} \times \dfrac{1}{4} \times \dfrac{2}{3} = \dfrac{4 \times 1 \times 1 \times 2}{7 \times 2 \times 4 \times 3} = \dfrac{8}{168} = \dfrac{1}{21}$

The above three examples were used to demonstrate the full process of multiplying common fractions. The canceling process was not used. However, the advantage of using the canceling process will be demonstrated by repeating the same three examples. Canceling will take place as soon and as often as possible. However, before demonstrating the advantage of using the canceling process on the previous problems, it is considered important that the process be thoroughly understood.

Canceling

Canceling is a valuable tool that can be used in the multiplication and division of fractions together with reducing a fraction to its lowest term. The process is also valuable when working with formulas.

The term *canceling* refers to the elimination of two identical numbers, or the dividing of a number in the numerator of a fraction and the denominator of a fraction by the same number. For example, examine the following multiplication of fractions problem.

$$\dfrac{(1)}{(3)} \; \dfrac{(3)}{(4)} \; \dfrac{(2)}{(7)}$$

First, change the 4 in the denominator to (2)(2)

Then, express the multiplication problem as:

$$\dfrac{(1)}{(3)} \; \dfrac{(3)}{(2)(2)} \; \dfrac{(2)}{(7)}$$

Now cancel out the prime numbers (one in the numerator and one in the denominator) that appear in both sections. In this example, cancel out the 3's and one of the two's. The completed operation will appear as follows:

$$\frac{\overset{\text{cancel}\;\downarrow\qquad\quad\downarrow\;\text{cancel}}{(1)\;(3)\;(2)}}{\underset{\text{cancel}\;\uparrow\qquad\uparrow\;\text{cancel}}{(3)\;(2)(2)\;(7)}}$$

The result appears as $\dfrac{1}{2 \times 7} = \dfrac{1}{14}$

Note: When the operation is complete after the canceling process, the result will be reduced to its lowest terms because there is no similar prime number that exists in both the numerator and the denominator.

Now let's look at the process with a larger multiplication problem.

$$\frac{(2)\;(7)\;(5)\;(13)\;(3)}{(3)\;(11)\;(7)\;(15)\;(3)}$$

Notice that after changing the 15 in the denominator to (3)(5), there is a 3, a 5, and a 7 in both the numerator and denominator. These can be directly canceled out. The solution would appear as follows.

$$\frac{(2)\;(7)\;(5)\;(13)\;(3)}{(3)\;(11)\;(7)\;(15)\;(3)} = \frac{(2)(7)(5)(13)(3)}{(3)(11)(7)(3)(5)(3)} = \frac{(2)(13)}{(3)(11)(3)}$$

Because there is no prime number remaining that is common to both the numerator and the denominator, the multiplication problem can be completed and the fraction obtained will be at its lowest term.

$$\frac{(2)\;(13)}{(3)\;(11)\;(3)} = \frac{26}{99}$$

The same process can be used when working with formulas. The restricting factor is that both sides of the equal sign have to be multipliers. No pluses or minuses are allowed. An acceptable problem would be as follows. For demonstrative purposes, this problem has been set up so that everything will be canceled out.

$$(2)(3)(5)(11)(13) = (5)(6)(11)(13)$$

Notice that the 5, 11, and 13 are common to both sides of the equal sign. By the canceling process, these numbers can be eliminated on both sides of the equal sign. The process and result appear as follows.

$$(2)(3)(5)(11)(13) = (5)(6)(11)(13)$$

What remains would appear as: $(2)(3) = (6)$. The 6 is a composite number consisting of the prime numbers 2 and 3, which makes the result appear as: $(2)(3) = (2)(3)$. The 2 and the 3 can be canceled on both sides of the equal sign.

$$(2)(3) = (2)(3)$$
$$0 = 0$$

This problem was completed in two steps. The process of changing the 6 to its prime numbers is normally done prior to any canceling. If this had been done, all of the numbers would have canceled out in the first step.

Now that the canceling process is understood, the previous three problems that were used to demonstrate the procedure for multiplying common fractions will be repeated, using the canceling process.

Problem: Multiply $\dfrac{1}{2} \times \dfrac{2}{3}$

Solution:
$$\dfrac{1}{2} \times \dfrac{2}{3} = \dfrac{(1)(2)}{(2)(3)}$$

cancel $------\downarrow$
$$\dfrac{(1)(2)}{(2)(3)} = \dfrac{1}{3}$$
cancel $--\uparrow$

Problem: Multiply $\dfrac{3}{4} \times \dfrac{2}{3} \times \dfrac{5}{7}$

Solution: $\dfrac{3}{4} \times \dfrac{2}{3} \times \dfrac{5}{7} = \dfrac{(3)(2)(5)}{(4)(3)(7)}$ change to $\dfrac{(3)(2)(5)}{(2)(2)(3)(7)}$

cancel $------\downarrow--\downarrow$
$$\dfrac{(3)(2)(5)}{(2)(2)(3)(7)} = \dfrac{5}{14}$$
cancel $------\uparrow--\uparrow$

Problem: Multiply $\dfrac{4}{7} \times \dfrac{1}{2} \times \dfrac{1}{4} \times \dfrac{2}{3}$

Solution: $\dfrac{4}{7} \times \dfrac{1}{2} \times \dfrac{1}{4} \times \dfrac{2}{3} = \dfrac{(4)(1)(1)(2)}{(7)(2)(4)(3)}$

cancel $-----\downarrow---------\downarrow$
$$\dfrac{(4)(1)(1)(2)}{(7)(2)(4)(3)} = \dfrac{1}{21}$$
cancel $---------\uparrow--\uparrow$

Multiplying Whole Numbers and Common Fractions

A whole number can be converted to an improper fraction by placing it over a 1. For example, a 3 can be converted to an improper fraction as follows: $3 = \frac{3}{1}$. Once a whole number is converted to an improper fraction, it can be directly entered into the procedure for multiplying common fractions. The full advantage of canceling will be included in the solution.

Problem: Multiply $4 \times \dfrac{3}{10}$

Solution: Set up the problem $\dfrac{4 \times 3}{1 \times 10}$

cancel $----------\downarrow$
$$\dfrac{(2)(2)(3)}{(1)(2)(5)} = \dfrac{6}{5} = 1\dfrac{1}{5}$$
cancel $----------\uparrow$

Problem: Multiply $4 \times \frac{7}{9} \times \frac{15}{20}$

Solution: Set up the problem $\frac{(2)(2)(7)(5)(3)}{(1)(3)(3)(2)(2)(5)}$

↓---↓-------↓---↓---------cancel

Change to $\frac{(2)(2)(7)(5)(3)}{(1)(3)(3)(2)(2)(5)} = \frac{7}{3}$

↑-------↑---↑---↑----cancel

Problem: Multiply $3 \times \frac{5}{7} \times \frac{2}{3} \times \frac{7}{11}$

Solution: Set up the problem $\frac{(3)\,(5)\,(2)\,(7)}{(1)\,(7)\,(3)\,(11)}$

↓-------------↓--------cancel

Change to $\frac{(3)\,(5)\,(2)\,(7)}{(1)\,(7)\,(3)\,(11)} = \frac{10}{11}$

↑---↑-------------cancel

Problem: Multiply $4 \times 6 \times \frac{3}{8} \times \frac{2}{9}$

Solution: Set up the problem $\frac{(4)\,(6)\,(3)\,(2)}{(1)\,(1)\,(8)\,(9)}$

↓---↓---↓---↓----↓-----------cancel

Change to $\frac{(2)\,(2)\,(2)\,(3)\,(3)\,(2)}{(1)\,(1)\,(2)\,(2)\,(2)\,(3)\,(3)} = 2$

↑---↑----↑---↑---↑------cancel

Multiply Mixed Numbers, Whole Numbers, and Common Fractions

To solve this type of multiplying fractions problem, it is first necessary to convert the mixed number to an improper fraction and the whole number to an improper fraction. The following problem can be used as an example.

Problem: Multiply $3\frac{2}{3} \times 5 \times \frac{3}{7} \times \frac{3}{5}$

Solution: Convert $3\frac{2}{3} = \frac{11}{3}$ Convert $5 = \frac{5}{1}$

Set up the problem $\frac{(11)\,(5)\,(3)\,(3)}{(3)\,(1)\,(7)\,(5)}$

cancel-----------↓---↓

$\frac{(11)\,(5)\,(3)\,(3)}{(3)\,(1)\,(7)\,(5)} = \frac{33}{7} = 4\frac{5}{7}$

cancel-----↑------------↑

Another example:

Problem: $4\frac{1}{2} \times 6\frac{1}{3} \times 5 \times \frac{7}{15}$

Solution: Convert $4\frac{1}{2} = \frac{9}{2}$ Convert $6\frac{1}{3} = \frac{19}{3}$

Set up the problem $\dfrac{(9)\,(19)\,(5)\,(7)}{(2)\,(3)\,(1)\,(15)}$

Change to $\dfrac{(3)\,(3)\,(19)\,(5)\,(7)}{(2)\,(3)\,(1)\,(3)\,(5)} = \dfrac{(19)\,(7)}{2} = \dfrac{133}{2} = 66\dfrac{1}{2}$

It's time to see whether you understood the procedures on how to multiply and cancel fractions. Let's start with the multiplication and canceling of original fractions and then extend the problems to multiplying whole numbers, common fractions, and then mixed numbers, whole numbers, and common fractions. Work each of the following problems. If you have any trouble with a problem, go back and review the procedures on how to work it. Use canceling whenever you can.

Problem 1. Multiply $\dfrac{5}{6} \times \dfrac{11}{12}$

Problem 2. Multiply $\dfrac{3}{4} \times \dfrac{7}{8} \times \dfrac{2}{3}$

Problem 3. Multiply $\dfrac{1}{2} \times \dfrac{3}{4} \times \dfrac{5}{6} \times \dfrac{12}{13}$

Problem 4. Multiply $7 \times \dfrac{11}{17}$

Problem 5. Multiply $15 \times \dfrac{3}{10} \times \dfrac{1}{2}$

Problem 6. Multiply $8 \times \dfrac{11}{24} \times \dfrac{5}{8} \times \dfrac{1}{2}$

Problem 7. Multiply $5 \times 19 \times 3\dfrac{1}{2} \times \dfrac{5}{6}$

Problem 8. Multiply $3 \times 5 \times 7\dfrac{5}{7} \times \dfrac{5}{45} \times \dfrac{3}{4}$

Reader's Practice Problems—5.4

Multiply $\dfrac{3}{5} \times \dfrac{2}{3}$

Multiply $\dfrac{2}{3} \times \dfrac{9}{11} \times \dfrac{5}{12}$

Multiply $4 \times \dfrac{2}{3} \times \dfrac{5}{12}$

Multiply $5 \times 2\dfrac{2}{3} \times 6\dfrac{1}{2} \times \dfrac{3}{5} \times \dfrac{2}{13}$

Multiply $\dfrac{7}{9} \times \dfrac{27}{35}$

Multiply $\dfrac{1}{3} \times \dfrac{3}{4} \times \dfrac{9}{11} \times \dfrac{7}{9}$

Multiply $7 \times 9 \times \dfrac{2}{21} \times \dfrac{5}{18}$

The answers are found in the Reader's Practice Problems section of the Appendix.

UNIT 4—REVIEW QUESTIONS

1. What is the general procedure to use for multiplying common fractions?
2. What are two concepts that should be remembered when multiplying two common fractions?
3. What makes the multiplication of common fractions different from the addition and subtraction processes?
4. What is the advantage of using canceling to determine the lowest term of a fraction?
5. What procedure should be used for multiplying mixed numbers, whole numbers, and common fractions?

TEST 5.4

Solve each of the following problems. Cancel when possible and reduce the fraction answer to its lowest term.

1. $\dfrac{2}{3} \times \dfrac{4}{7}$
2. $\dfrac{3}{4} \times \dfrac{5}{7}$
3. $\dfrac{2}{3} \times \dfrac{1}{5} \times \dfrac{4}{7}$
4. $\dfrac{1}{4} \times \dfrac{1}{2} \times \dfrac{2}{9}$
5. $\dfrac{1}{2} \times \dfrac{3}{5} \times \dfrac{2}{3} \times \dfrac{11}{13}$
6. $\dfrac{2}{9} \times \dfrac{2}{3} \times \dfrac{3}{4} \times \dfrac{1}{2}$
7. $5 \times \dfrac{1}{2} \times \dfrac{2}{3} \times \dfrac{2}{5}$
8. $4 \times \dfrac{3}{4} \times \dfrac{5}{7} \times \dfrac{7}{11}$
9. $4\dfrac{1}{2} \times \dfrac{3}{5} \times \dfrac{2}{3} \times \dfrac{7}{11}$
10. $2\dfrac{3}{4} \times \dfrac{1}{2} \times \dfrac{4}{7} \times \dfrac{2}{5}$

11. This situation is a practical approach to a real life situation. It actually happened a few years ago in a fire station in a large city on the West coast. Here is the problem and the solution.

Problem: When it comes to desserts in a fire station, every firefighter wants to make sure that no other firefighter gets a larger share than he or she. Seven firefighters were on duty the day that the cook baked a delicious pumpkin pie (Figure 5.10). The cook stared at the cooling pie, wondering how he was going to cut it into seven equal pieces. Then an idea dawned on him that would solve the problem and make every firefighter happy.

Figure 5.10

A freshly baked pie

Figure 5.11

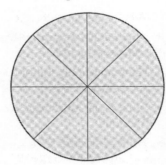

Solution: It was easy to cut the pie into eight equal pieces, which he did (Figure 5.11). He then threw one of the pieces into the trash. The solution pleased all of the firefighters. No firefighter received a larger portion than any other firefighter.

Problem for the reader: How much of their share of the pie did each firefighter lose?

SOLUTION HELP—Each firefighter lost $\frac{1}{7}$ of $\frac{1}{8}$ of the pie. To solve the problem, multiply $\frac{1}{7} \times \frac{1}{8}$

12. The volume of a rectangular container can be determined by the formula

$$\text{Volume} = lwh$$

where   l = length
w = width
h = height

Figure 5.12

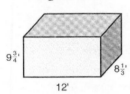

What is the volume of the rectangular container shown in Figure 5.12?

SOLUTION HELP—To solve the volume, multiply the length (12 feet) × the width ($8\frac{1}{3}$ feet) × the height ($9\frac{3}{4}$ feet)

13. The Topnotch Fire Department reported that they responded to 1,255 alarms last year, $\frac{4}{5}$ of which were for medical emergencies. Of the medical emergency responses, $\frac{1}{8}$ were for injuries involving vehicles similar to the one shown in Figure 5.13. How many responses during the year were for this type of medical emergency? Round the answer to the nearest whole number.

SOLUTION HELP—First, multiply the 1,255 (alarms) by $\frac{4}{5}$ (medical emergencies) Next, multiply the answer by $\frac{1}{8}$. Round off the response.

14. In 2007, the Big City Fire Department received a total of 600 false alarms. Of these, $\frac{1}{2}$ were in Division 3. Of those in Division 3, $\frac{1}{3}$ were in Battalion 2. Of those in Battalion 2, $\frac{3}{4}$ were in Engine 7's first-in district. How many were in Division 3, in Battalion 2, and in Engine 7's district?

SOLUTION HELP—First, multiply $\frac{1}{2} \times$ 600 (Division 3). Second, multiply $\frac{1}{3} \times$ the result (Battalion 2). Next, multiply $\frac{3}{4} \times$ the result (Engine 7).

15. During the month of August, Engine 5 responded to 120 calls for medical assistance: $\frac{1}{3}$ of the calls were for heart attacks, $\frac{1}{5}$ for difficult breathing, and $\frac{1}{4}$ for vehicle accidents. Of the 120 responses, how many were for either a heart attack or difficult breathing?

SOLUTION HELP—First, multiply $\frac{1}{3} \times$ 120. Second, multiply $\frac{1}{5} \times$ 120. Then add the two results of the multiplications.

Figure 5.13

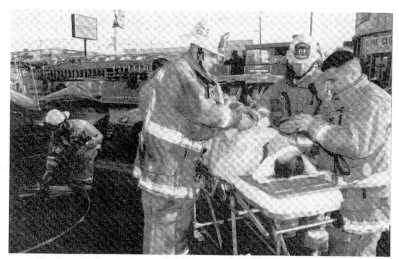

Courtesy of Rick McClure, LAFD

16. The formula for determining the area of a rectangle is: area = length × width. What would be the area of the first floor of a small structure that has a length of $65\frac{1}{2}$ feet and a width of $34\frac{1}{3}$ feet?

    SOLUTION HELP—Multiply the length × the width.

17. During a seven-day period, a paramedic running out of the ambulance housed in Engine 7's quarters worked $8\frac{1}{2}$ hours on three days, $9\frac{1}{4}$ hours on two days, and $9\frac{1}{3}$ hours on two days. How many total hours did she work during the seven days?

    SOLUTION HELP—Multiply $3 \times 8\frac{1}{2} + 2 \times 9\frac{1}{4} + 2 \times 9\frac{1}{3}$

18. Most fire department firefighters work a 24-hour shift. During one of these shifts, the members of the A Platoon of Engine 2 spent $\frac{1}{12}$ of their time in training, $\frac{1}{8}$ on fire prevention inspections, $\frac{1}{12}$ doing housework, and $\frac{1}{24}$ on physical fitness exercises. How much time did the members devote to the four activities during the 24-hour shift?

    SOLUTION HELP—Add $\frac{1}{12} + \frac{1}{8} + \frac{1}{12} + \frac{1}{24}$

19. The amount of pressure (psi) created by a column of water is approximately $\frac{11}{25}$ of a pound for each feet of head. If the head is 72 feet, approximately how much pressure would it create?

    SOLUTION HELP—Multiply $\frac{11}{25} \times 72$

20. A small fire occurred in a bedroom of a single-family dwelling. The room measured $9\frac{1}{2}$ feet by $11\frac{3}{4}$ feet. What was the area of the bedroom?

    SOLUTION HELP—Multiply $9\frac{1}{2} \times 11\frac{3}{4}$

21. The formula used to determine the volume of a cylindrical container is:

    $$V = .7854\, D^2 H$$

    where   D = diameter
                    H = height

What would be the volume of a cylindrical container that has a diameter of 5 feet and is 9 feet in height?

SOLUTION HELP—Multiply .7854 × 5 × 5 × 9

22. The formula for determining the velocity flow is:

$$V = 8\sqrt{H}$$

where   H = head

What would be the velocity flow from a hole in the side of a tank that is located 50 feet below the level of the water in the tank? The square root of 50 is 7.07.

SOLUTION HELP—Multiply 8 × 7.07

23. The formula for determining the nozzle reaction from a tip is:

$$NR = 1.57\, D^2 P$$

where   D = diameter of the tip

P = nozzle pressure

What would be the nozzle reaction from a $1\frac{1}{2}$-inch tip when the nozzle pressure is 40 psi?

SOLUTION HELP—Multiply 1.57 × 1.5 × 1.5 × 40

24. The weight capacity for water is determined by the formula:

$$\text{Capacity} = 62.5\, V$$

where   V = volume

What would be the weight capacity of a container when the volume of the container is 15 cubic feet?

SOLUTION HELP—Multiply 62.5 × 15

25. The gallon capacity of water is determined by the formula:

$$\text{Gallons} = 7.48\, V$$

where   V = volume

What would be the gallon capacity of a container that has a volume of 110 cubic feet?

SOLUTION HELP—Multiply 7.48 × 110

UNIT 5—DIVISION OF FRACTIONS

Solving a division of fractions problem is almost identical to solving a multiplication of fractions problem. The only difference is that before starting the multiplication process, it is necessary to reverse the positions of the numerator and denominator of the divisor. For example, divide $\frac{1}{4}$ by $\frac{1}{2}$. This would probably be written as $\frac{1}{4} \div \frac{1}{2}$. The solution would appear as follows.

$$\frac{1}{4} \div \frac{1}{2} \text{ reverse divisor} \quad \frac{1}{4} \times \frac{2}{1} \quad \text{multiply} = \frac{2}{4} = \frac{1}{2}$$

Problem: Divide $\frac{7}{8}$ by $\frac{3}{8}$

Solution:

$$\frac{7}{8} \div \frac{3}{8}$$

reverse divisor
$$\frac{7}{8} \times \frac{8}{3}$$

multiply
$$\frac{(7)(8)}{(8)(3)}$$

cancel --------↓
$$\frac{(7)(8)}{(8)(3)} = \frac{7}{3} = 2\frac{1}{3}$$
cancel -----↑

Dividing a Whole Number by a Mixed Number

The first step when dividing a whole number by a mixed number is to convert both the whole number and the mixed number to improper fractions. An example is to divide 5 by $2\frac{2}{3}$.

Convert the 5 to an improper fraction: $\quad 5 = \frac{5}{1}$

Convert the $2\frac{2}{3}$ to an improper fraction: $\quad 2\frac{2}{3} = \frac{8}{3}$

Set up the problem: $\quad \frac{5}{1} \div \frac{8}{3}$

Reverse the divisor: $\quad \frac{5}{1} \times \frac{3}{8}$

Solve the problem: $\quad \frac{5}{1} \times \frac{3}{8} = \frac{15}{8} = 1\frac{7}{8}$

Dividing a Mixed Number by a Mixed Number

The first step when dividing a mixed number by a mixed number is to convert both numbers to improper fractions. For example, divide $3\frac{1}{2}$ by $2\frac{2}{5}$.

Convert $3\frac{1}{2}$ to an improper fraction: $\quad 3\frac{1}{2} = \frac{7}{2}$

Convert $2\frac{2}{5}$ to an improper fraction: $\quad 2\frac{2}{5} = \frac{12}{5}$

Set up the problem: $\quad \frac{7}{2} \div \frac{12}{5}$

Reverse the divisor: $\quad \frac{7}{2} \times \frac{5}{12}$

Solve the problem: $\quad \frac{7}{2} \times \frac{5}{12} = \frac{35}{24} = 1\frac{11}{24}$

As a final check prior to working the last Reader's Practice Problems of this chapter to see whether you know how to divide fractions, work each of the following problems. If you have any difficulty with any of them, go back to the section on dividing fractions and review it again.

Problem 1. Divide $\frac{5}{7}$ by $\frac{3}{4}$

Problem 2. Divide $\frac{3}{4}$ by $\frac{3}{10}$

Problem 3. Divide 11 by $6\frac{1}{3}$

Problem 4. Divide 6 by $3\frac{4}{5}$

Problem 5. Divide $5\frac{1}{2}$ by $6\frac{7}{8}$

Problem 6. Divide $8\frac{1}{3}$ by $5\frac{3}{4}$

Reader's Practice Problems—5.5

Divide $\frac{4}{5}$ by $\frac{2}{3}$

Divide $\frac{7}{8}$ by $\frac{5}{7}$

Divide $\frac{4}{5}$ by 7

Divide 8 by $1\frac{2}{3}$

Divide $6\frac{5}{6}$ by $4\frac{1}{3}$

Divide $2\frac{3}{5}$ by $1\frac{2}{7}$

The answers are found in the Reader's Practice Problems section of the Appendix.

UNIT 5—REVIEW QUESTIONS

1. What is the primary difference between the multiplication procedure and the division procedure for common fractions?
2. What is the first step when dividing a whole number by a mixed number?
3. What is the first step when dividing a mixed number by a mixed number?

TEST 5.5

1. $\frac{2}{5} \div \frac{3}{4}$
2. $\frac{5}{9} \div \frac{2}{7}$
3. $5 \div 1\frac{1}{3}$
4. $8 \div 3\frac{1}{3}$
5. $1\frac{3}{4} \div 2\frac{1}{5}$
6. $3\frac{1}{7} \div 2\frac{1}{5}$
7. $\frac{4}{9} \div 1\frac{1}{3}$
8. $8 \div 3\frac{2}{3}$
9. $5\frac{1}{2} \div 1\frac{5}{6}$
10. $12\frac{1}{5} \div 3\frac{3}{4}$

11. During a six-day period, a paramedic assigned to an ambulance housed in Engine 5's station worked a total of $56\frac{3}{4}$ hours. If he worked the same number of hours each day, how many hours a day did he work?

    SOLUTION HELP—Divide $56\frac{3}{4}$ by 6

12. Engine 5 responded to an incident where gasoline was leaking out of a hole in the side of a 1,000-gallon tanker truck. It was estimated that the gasoline was leaking out of the hole at a rate of $8\frac{1}{3}$ gallons per minute (gpm). At this rate, how long will it take for all of the gasoline to leak out of the tanker truck?

    SOLUTION HELP—Divide 1000 by $8\frac{1}{3}$

13. Water weighs approximately $62\frac{1}{2}$ pounds per cubic feet. A gallon of water weighs approximately $8\frac{1}{3}$ pounds. Approximately how many gallons of water are in a cubic feet of water?

    SOLUTION HELP—Divide $62\frac{1}{2}$ by $8\frac{1}{3}$

14. A swimming pool holds 30,000 gallons of water. A pumper is drafting from the pool and discharging water through a hose line at the rate of 180 gpm. Assuming all the water can be removed from the pool, how much time would it take to empty the pool?

    SOLUTION HELP—Divide 30,000 by 180

15. A water tank on an apparatus contains approximately 1,500 cubic feet of water. There are approximately $7\frac{1}{2}$ gallons in a cubic feet of water. How many gallons of water will the tank hold?

    SOLUTION HELP—Divide 1,500 by $7\frac{1}{2}$

16. Engine 12 responds to approximately $20\frac{2}{3}$ calls per month for injuries in vehicle accidents. Approximately how long will it take the company to respond to 152 calls?

    SOLUTION HELP—Divide 152 by $20\frac{2}{3}$

17. A 5,000-gallon water tanker crew stops at a hydrant to fill up the empty tanker. They connect to the hydrant with a $2\frac{1}{2}$-inch line that can fill the tanker at a rate of $180\frac{2}{3}$ gpm. How long will it take to fill the tanker?

    SOLUTION HELP—Divide 5,000 by $180\frac{2}{3}$

18. A high-rise building is 380 feet in height. There are $9\frac{1}{2}$ feet per floor. How many floors are there in the building?

    SOLUTION HELP—Divide 380 by $9\frac{1}{2}$

19. A 2-inch tip discharges approximately 1,000 gpm. How many tips discharging $165\frac{2}{3}$ gpm each will it take to equal the same flow as the 2-inch tip?

    SOLUTION HELP—Divide 1,000 by $165\frac{2}{3}$

20. $7 \div 4\frac{1}{3}$ 21. $12 \div 5\frac{3}{5}$ 22. $3\frac{5}{6} \div 7\frac{3}{4}$

23. $5\frac{2}{3} \div 7\frac{1}{2}$ 24. $7\frac{1}{5} \div 5\frac{1}{2}$

25. The volume of a rectangular container is determined by the formula:

    $$V = l \times w \times h$$

    where   l = length

    w = width

    h = height

    If the length of a rectangular container is 10 feet, the width is 6 feet, and the volume is 480 cubic feet, what is the height?

    SOLUTION HELP—Multiply 10×6 and divide the result into 480.

26. The gallons of water that a container can hold is determined by the formula:

    $$\text{Gallons} = 7.48\, V$$

    where   V = volume of the container

    If the volume of a container is 388.96 cubic feet, how many gallons of water can it hold?

    SOLUTION HELP—Multiply 388.96×7.48

27. The momentary nozzle reaction can be determined by the formula:

$$MNR = 1.88D^2P$$

where D = the diameter of the tip being used
       P = nozzle pressure

If the tip size is $1\frac{1}{4}$ and the MNR is 161.56 pounds, what would be the approximate nozzle pressure?

SOLUTION HELP—Multiply 1.88 × 1.25 × 1.25 and divide the result into 161.56.

28. The theoretical height to which water will rise when drafting can be determined by the formula:

$$height = 1.13\, Hg$$

where Hg = inches of mercury

If the theoretical height is 24.86, how many inches of mercury does the suction gauge indicate?

SOLUTION HELP—Divide 24.86 by 1.13

29. The pressure reduction (Hg) needed in order to lift water a given amount when drafting can be determined by the formula:

$$Hg = .885\, h$$

where h = height

If the pressure reduction needed to lift water is 15 inches, how high could water be lifted?

SOLUTION HELP—Divide 15 by .885

30. Use the formula

$$Area = .7854\, D^2$$

to determine the area of a circle. If the area of a circle is 3.1416 inches, what is the diameter of the circle?

SOLUTION HELP—Divide 3.1416 by .7854 and take $\frac{1}{2}$ of the answer.

Chapter Summary

This chapter was divided into five units. The first unit explained the meaning of working with fractions. This explanation included the definition of words that are used in the other four units, an introduction to the various types of common fractions, and a presentation of the concepts of whole numbers with an explanation of the term "composite number." This unit also included the methods for changing a fraction to an equivalent fraction, how to convert a mixed number to an improper fraction, and the method for converting an improper fraction to a mixed number. Problems for the reader to solve and a small section on review questions were also included.

The other four units were divided into the four procedures for working with fractions. Unit 2 was devoted to addition problems, Unit 3 to subtraction problems, Unit 4 to multiplication problems, and Unit 5 to division problems. Each unit included problems for the reader to solve and a section on review questions.

CHAPTER 6
WORKING WITH DECIMALS

KEY TERMS

decimal number
dividend
divisor

quotient
repeating decimal

rounding a number
terminating decimal

OBJECTIVES AND GOALS

Upon completing this chapter, the reader should be able to:

- Explain the three parts of a decimal fraction.
- Define the terms decimal number, terminating decimal, and repeating decimal.
- Explain the procedure for rounding a number.
- Convert a fraction to a decimal.
- Convert a decimal to a fraction.
- Explain the difference between adding whole numbers and adding decimals.
- Explain the procedure for locating a decimal point in an addition problem.
- Prove that the answers to addition, subtraction, multiplication, and division decimal problems are correct using the "excess of nines" method.
- Explain the difference between subtracting whole numbers and subtracting decimals.
- Describe the procedure for setting up a subtraction of decimals problem.
- Describe the difference between multiplying decimals and multiplying whole numbers.
- Describe the procedure for multiplying decimals.
- Explain the two primary differences between working with division problems involving whole numbers and those involving decimals.

UNIT 1

INTRODUCTION

Like fractions, decimals are a portion of a whole. Decimals are as much a part of an individual's daily life as are whole numbers. In fact, in many situations they are more than a part. Almost everything an individual buys has a portion of its cost portrayed as a decimal.

Seldom is a sales price of a piece of clothing, an item of jewelry, or any other major item listed as a whole number. Prices such as $5.99 or $14.99 are common. Additionally, statistics on cost of living, employment, and other such information is usually given as "a rise of 4.7 percent, or down 0.6 percent."

The portion of the number 1 that is illustrated by a decimal point can be compared with the same portion illustrated by a fraction. For example, in the number 10.325, the decimal, .325, refers to $\frac{3}{8}$ of the number 1, or decimal .375 of the number 1.

Fraction	Decimal
$\frac{1}{8}$	.125
$\frac{1}{4}$	.25
$\frac{3}{8}$	.375
$\frac{1}{2}$	.5
$\frac{5}{8}$	.625
$\frac{3}{4}$	.75
$\frac{7}{8}$	.875

Decimal fractions are represented in multiple units of 10. Combined with a whole number, a decimal fraction is identified in three parts: the whole number, the decimal point, and the decimal fraction. The identification of each portion can be illustrated by its location within the following placement chart.

WHOLE NUMBER					DECIMAL POINT	DECIMAL FRACTION			
2	3	4	5	6	.	3	5	7	8
ten thousands place	thousands place	hundreds place	tens place	units place	decimal place	tenths place	hundredths place	thousandths place	ten thousandths place
10,000	1000	100	10	1		1/10	1/100	1/1000	1/10,000

WORDS TO REMEMBER

decimal number—A number made up of a whole number and a decimal fraction. For example, in the decimal number 6.27, 6 is the whole number and .27 is the decimal fraction.

terminating decimal—A decimal whose decimal fraction stops. For example, 0.125 is a terminating decimal.

repeating decimal—A decimal whose decimal fraction repeats itself forever. For example, the decimal fraction 0.142857142857142857 is a repeating decimal.

Rounding Numbers

Rounding a number refers to converting a number to an approximation of that number rather than using the real number. It is normally used by administrative personnel of a fire department and to present statistical data to others who are not necessarily interested in the exact number. For example, if the budget for the Fire Department for a community is $8,452,178.55, the budget might be presented to the public as $8,500,000, or even released to the press as $8\frac{1}{2}$ million dollars.

At other times, instructions are given to a provider to round a number to a particular place value. For example, instruction may be given to round a number to the hundredths place value. This place value is referred to as the second point after the decimal for many people who are not familiar with the place value chart previously illustrated in this chapter. Rounding an answer to the second point after the decimal is also a common procedure that is used in many mathematical examinations.

PROCEDURE FOR ROUNDING A NUMBER

To round a number to a particular place value, look at the digit to the right of the place value. If this digit is less than 5, everything to the right of the place value can be dropped or changed to zeros. If this digit is 5 or larger, the place value should be increased by 1 and everything after the place value reduced to zeros. For example:

Problem: Round 56,319 to the nearest hundreds.

Solution: The 3 is in the hundreds place value. The digit to the right of the 3 is less than 5. Therefore, the 1 and the 9 should be changed to zeros. The answer to the problem would be 56,300. Following is the example of how this developed.

```
                              place value
                                   ↓
        original number        56,319
                                  ↑↑------change to zeros
        answer                 56,300
```

Problem: Round 1,623,476,202 to the 10,000 place value.

Solution: The 7 is in the 10,000 place value. The number to the right of the 7 is 6. Because the 6 is larger than 5, the place value should be increased by 1 and all the digits after the increased place value should be changed to zero. Following is the example of how this developed.

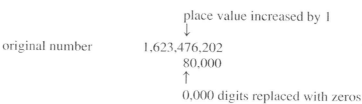

Answer: 1,623,480,000

Problem: The budget for the Dow City Fire Department was adopted by the City Council as $105,563,478,002.87. The Fire Chief decided that he would list the budget in the local paper to the nearest $100,000. Round off the budget to show the figure that he released to the local paper.

Solution: The adopted budget was $105,563,478,002.87. The 4 of this budget is located at the 100,000 place value. The digit to the right of the place value is 7. Because this is 5 or larger, the place value should be increased by 1, and all the digits to the right of the place value, including the 7, should be changed to zero, and the digits after the decimal point dropped. This is developed as follows.

$$\text{adopted budget} \quad \$105{,}563{,}\overset{\downarrow \text{place value increased by one}}{5}78{,}002.87 \; \leftarrow \; \leftarrow 8 \text{ \& } 7 \text{ dropped}$$

$$\underset{\text{changed to zeros}}{\underset{00\;000}{\uparrow\uparrow\;\uparrow\uparrow\uparrow}}$$

Answer: $105,563,500,000

Converting a Fraction to a Decimal

To convert a fraction to a decimal, divide the numerator by the denominator. For example, to convert $\frac{3}{7}$ to a decimal, the process would be as follows.

$$\frac{3}{7} \rightarrow 7\overline{\smash{)}3.0000} = 7\overline{\smash{)}3.000000}^{\,0.428571}$$

Note: 0.428571 is a repeating decimal. A repeating decimal is one whose decimal fraction repeats itself forever.

Rounding it to the second decimal point, it would appear as 0.43.

Problem: Convert the fraction $\frac{5}{8}$ to a decimal.

Solution: $\frac{5}{8} \rightarrow 8\overline{\smash{)}5.0000} = 8\overline{\smash{)}5.0000}^{\,.625}$

Note: .625 is a terminating decimal.

Rounding it to the second decimal point, it would appear as .63.

Converting a Decimal to a Fraction

A decimal can be converted to a fraction by first placing the decimal over 1. Then multiply both the numerator and the denominator repeatedly by 10 until the numerator is a whole number. Following is a demonstration of this process.

Problem: Convert 0.36 to a fraction.

Solution: $0.36 = \dfrac{0.36}{1} = \dfrac{(0.36)(10)}{(1)(10)} = \dfrac{3.6}{10} = \dfrac{(3.6)(10)}{(10)(10)} = \dfrac{36}{100}$

Convert $\frac{36}{100}$ to its lowest term by dividing both the numerator and denominator by 4.

$$36 \div 4 = 9$$
$$100 \div 4 = 25$$

Then $\frac{36}{100}$ reduced to its lowest term $= \frac{9}{25}$

Problem: Convert .72 to a fraction.

Solution: $72 = \frac{72}{1} = \frac{(.72)(10)}{(1)(10)} = \frac{7.2}{10} = \frac{(7.2)(10)}{(10)(10)} = \frac{72}{100}$

Convert the fraction $\frac{72}{100}$ to its lowest terms by dividing both the numerator and the denominator by 4. $\frac{72}{100} = \frac{18}{25}$

To see whether you understood all of the previous examples, solve the following problems and place your answer in the space provided. Be sure to reduce all fractions to their lowest terms. Feel free to go back to the example to review how to solve it.

Problem 1: Round 1,453,947 to the nearest hundreds.

Solution: 1,453,947 rounded to the nearest hundreds = _____

Problem 2: The City Council of Park City adopted the budget for the Fire Department at $67,456,753.77. The fire chief decided to release the budget to the local media as $67,457,000. To what place value did the chief release the adopted budget to the local press?

Solution: The budget was released to the press at the _____ place value.

Problem 3: Convert the fraction $\frac{3}{7}$ to a decimal. Round to the second decimal point.

Solution: $\frac{3}{7}$ converted to a decimal = _____

Problem 4: Convert the decimal .467 to a fraction. Be sure to reduce the fraction to its lowest terms.

Solution: .464 converted to a fraction = _____.

Reader's Practice Problems—6.1

1. Round 5,476.6547 to the nearest hundredths place.
2. Round 2,535.7462 to the nearest hundreds place.
3. Round 42.75771 to the second decimal point.
4. Convert $\frac{7}{8}$ to a decimal. Round off to third decimal point.
5. Convert $\frac{3}{7}$ to a decimal. Round off to third decimal point.
6. Convert .365 to a fraction.
7. Convert .19 to a fraction.

The answers are found in the Reader's Practice Problems section of the Appendix.

UNIT 1—REVIEW QUESTIONS

1. Explain the three parts of a decimal fraction.
2. Define decimal number, terminating decimal, and repeating decimal.
3. Explain the procedure for rounding a number.
4. Explain the procedure for rounding a decimal.

5. Explain how to convert a fraction to a decimal.
6. Explain how to convert a decimal to a fraction.

TEST 6.1

1. Round 8,756 to the nearest hundreds.
2. Round 27.5846 to the nearest hundredth.
3. Convert $\frac{5}{11}$ to a decimal. Provide an answer to the nearest thousandths.
4. Convert 0.17 to a fraction.
5. Round 824,449 to the nearest hundreds.
6. Round 78.4249 to the second decimal point.
7. Convert $\frac{4}{7}$ to a decimal. Provide an answer to the nearest tenth.
8. Convert $\frac{5}{9}$ to a decimal. Provide an answer to the second decimal point.
9. Convert 0.23 to a fraction.
10. Convert 0.28 to a fraction. Reduce the fraction to its lowest term.
11. Convert $\frac{3}{4}$ to a decimal.
12. Convert 0.66 to a fraction.
13. Convert $\frac{69}{100}$ to a decimal.
14. Convert $\frac{5}{6}$ to a decimal.
15. Convert 0.27 to a fraction.

UNIT 2—ADDITION OF DECIMALS

Principles of Adding Decimals

The only difference between adding whole numbers and adding decimals is that the decimals in the numbers to be added should be lined up. It is also good practice to add zeros to any decimal that does not have the same number of digits in the decimal as one of those to be added. The proper procedure to follow is:

1. Write the figures to be added vertically in the same manner as if adding whole numbers. Line up the decimal points.
2. If there are any empty place-value positions, add zeros.
3. Complete the addition problem.
4. Round off the answer, if required.

Problem: Add 254.671 + 96.34 + 47.67. Provide an answer to the nearest hundredth.

Solution:

$$\begin{array}{r} 254.671 \\ 96.340 \\ 47.670 \\ \hline 398.681 \end{array}$$

Answer: 398.68—rounding off to the nearest hundredth

Problem:

$$\begin{array}{r} 77.43 \\ 287.98 \\ 934.347 \\ 125.4 \\ \hline \end{array}$$

Solution: First add the necessary zeros, then complete the addition.

$$
\begin{array}{r}
77.430 \\
287.980 \\
934.347 \\
\underline{125.400} \\
1{,}425.157
\end{array}
$$

Proving Decimal Addition Problems

Prove that the answer is correct by the "excess of nines" method. When proving the answer, prove it prior to rounding off the answer, otherwise it will not check out. When proving a decimal addition problem, ignore the decimal points.

The procedure for checking an addition problem was introduced in Chapter 2. It is repeated here as a review, using the previous two problems.

Problem 1:

$$
\begin{array}{r}
254.671 \\
96.340 \\
\underline{47.670} \\
398.681
\end{array}
$$

Step 1: Determine the "excess of nines" of each of the addends.

$$254.671 = 2 + 5 + 4 + 6 + 7 + 1 = 25 \quad 2 + 5 = 7$$
$$96.340 = 9 + 6 + 3 + 4 + 0 = 22 \quad 2 + 2 = 4$$
$$47.670 = 4 + 7 + 6 + 7 + 0 = 24 \quad 2 + 4 = 6$$

Step 2: Determine the "excess of nines" of the "excess of nines" of the addends.

$$7 + 4 + 6 = 17 \quad 1 + 7 = 8 \leftarrow \leftarrow \leftarrow \leftarrow \text{same} \leftarrow \leftarrow \leftarrow \leftarrow$$

Step 3: Determine the "excess of nines" of the answer.

$$398.681 = 3 + 9 + 8 + 6 + 8 + 1 = 35 \quad 3 + 5 = 8 \leftarrow \leftarrow \leftarrow$$

Conclusion: Because the "excess of nines" of the addends' "excess of nines" and the "excess of nines" of the answer are the same, it can be assumed that the problem has been worked correctly.

To complete the problem, round off the answer.

$$398.681 = 398.68$$

Problem 2:

$$
\begin{array}{rr}
77.430 & -3 \\
287.980 & -7 \\
934.347 & -3 \\
\underline{125.400} & -3 \\
1{,}425.157 &
\end{array}
$$

Step 1: Determine the "excess of nines" of each of the addends.

$$77.430 = 7 + 7 + 4 + 3 + 0 = 21 \quad 2 + 1 = 3$$
$$287.980 = 2 + 8 + 7 + 9 + 8 + 0 = 34 \quad 3 + 4 = 7$$
$$934.347 = 9 + 3 + 4 + 3 + 4 + 7 = 30 \quad 3 + 0 = 3$$
$$125.400 = 1 + 2 + 5 + 4 + 0 + 0 = 12 \quad 1 + 2 = 3$$

Step 2: Determine the "excess of nines" of the "excess of nines" of the addends.

$3 + 7 + 3 + 3 = 16 \quad 1 + 6 = 7 \leftarrow \leftarrow \leftarrow \text{same} \leftarrow \leftarrow \leftarrow \leftarrow \leftarrow \leftarrow$

Step 3: Determine the "excess of nines" of the answer.

$1{,}425.157 = 1 + 4 + 2 + 5 + 1 + 5 + 7 = 25 \quad 2 + 5 = 7 \rightarrow \rightarrow \rightarrow \uparrow$

Conclusion: Because the "excess of nines" of the addends' "excess of nines" and the "excess of nines" of the answer are the same, it can be assumed that the problem has been worked correctly.

Following is a suggestion for setting up the three steps when working the problem.

```
   77.430   − 3
  287.980   − 7                3 + 7 + 3 + 3 = 16
  934.347   − 3                        1 + 6 = 7 ← ← ← ← ←↑
  125.400   − 3                                              same
1,425.157              1 + 4 + 2 + 5 + 1 + 5 + 7 = 25         ↑
                                         2 + 5 = 7 ← ← ← ←
```

A PRACTICAL SOLUTION WHEN ADDING DECIMALS

When you deposit a check in your checking account, in essence you are adding decimals to determine your new balance. There is a method you should be aware of that can occasionally be used that simplifies the addition.

Following is an example of this method.

You have a balance of $6,723.55 in your account. You deposit a check in the account that amounts to $29.97. Rather than complete the addition, add $30 to the account, and subtract 3 cents from what was in there. The account would now total $6,753.52. Look for the opportunities to use this system. It makes your work much easier and less confusing.

Reader's Practice Problems—6.2

Add the following problems. Round off the answers to the second decimal point. Prove each of the answers by the "excess of nines" method.

1. Add 345.65 + 237.789 + 45.78 + 734.831
2. Add 9843.67 + 4532.78 + 8743.9

3.	45.747	4.	763.89	5.	3487.265
	67.3465		16.67		1276.4
	4.784		397.7		7764.958
	98.99		7.482		897.21

The answers are in the Reader's Practice Problems section of the Appendix.

UNIT 2—REVIEW QUESTIONS

1. What is the three-step procedure for solving addition decimal problems?
2. What procedure is used for proving that an addition decimal problem has been worked correctly?

TEST 6.2

Solve the following problems and prove your answers. Round all answers to the nearest hundredth.

1. Add 34.72 + 46.73
2. Add 197.62 + 341.294
3. Add 342.17 + 87.453 + 69.347
4. Add 974.457 + 342.754 + 777.684
5. Add 4 + 2.7 + 32.45
6. Add 33.33 + 1.43 + 0.52
7. Add 1021.76 + 2.74 + 71.23
8. Add 2.45 + 3.46 + 13.23 + 45.62
9. Add 123.468 + 453.24 + 2343.34
10. Add 999.98 + 4456.34 + 77.342

Note to the reader: Many of the word problems in this test and in the other tests in this chapter are not true-life situations. The reason for this is that the problems are given using decimals whereas similar true-life problems would most likely be given in whole numbers. The change has been made to provide the reader with the experience of working with decimal word problems.

11. The balance in your checkbook is $3,452.76. You deposit a check with an amount of $43.97. What is the total you now have in your account?

    SOLUTION HELP—Add $44 to your account of $3,452.76 and subtract .03 from your checkbook.

12. The pumper shown in Figure 6.1 is supplying water to the three lines shown in the figure. What is the total amount of water being discharged from the pumper?

    SOLUTION HELP—Add the flow from the three lines.

Figure 6.1

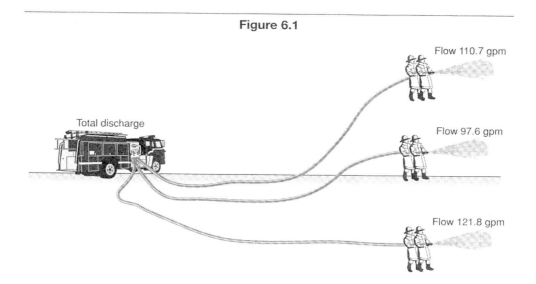

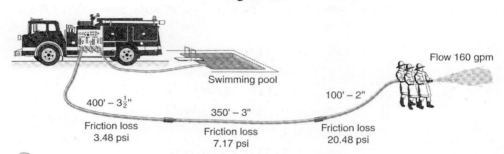

Figure 6.2

13. What is the total friction loss in the line leading from the pumper to the firefighters holding it in Figure 6.2?

    SOLUTION HELP—Add the three friction losses shown in the drawing.

14. Engine 5's crew spent the following amount of time on in-house training during the month of November: first shift—2.3 hours, second shift—1.9 hours, third shift—2.5 hours, fourth shift—1.8 hours, fifth shift—2.2 hours. How much time did they spend on in-house training during the month of November?

    SOLUTION HELP—Add the hours for the five shifts.

15. On the first six responses during the month of February, Engine 2 traveled the following distances: first—4.3 miles, second—1.8 miles, third—6.2 miles, fourth—5.4 miles, fifth—7.1 miles, sixth—3.7 miles. How many miles did the company travel on these six responses?

    SOLUTION HELP—Add the six mileage amounts.

16. Engine 9 responded to a vehicle accident where three individuals were injured. The firefighter EMTs immediately took the vital signs on all three. Victim A had a blood pressure of 128 over 79 and a body temperature of 98. Victim B had a blood pressure of 136 over 82 and a body temperature of 99. Victim C had a blood pressure of 143 over 87 and a body temperature of 102. What was the combined systolic blood pressure of the three victims?

    SOLUTION HELP—Add 128 + 136 + 143

17. Using the information in Problem 16, what was the combined body temperatures of the three victims?

    SOLUTION HELP—Add 98 + 99 + 102

18. Using the information in Problem 16, what was the combined systolic and diastolic blood pressures of victims B and C?

    SOLUTION HELP—Add 136 + 82 + 143 + 87

19. Using the information in Figure 6.3, what is the total flow from the three nozzle tips?

    SOLUTION HELP—Add 177.94 + 259.18 + 212.33

20. Using the information in Figure 6.3, what is the RPDP for line C?

    The formula for determining the RPDP is:

    $$RPDP = FL + NP$$

    where   FL = friction loss in the hose line
    NP = nozzle pressure

    SOLUTION HELP—Add 59.93 + 51.94

Figure 6.3

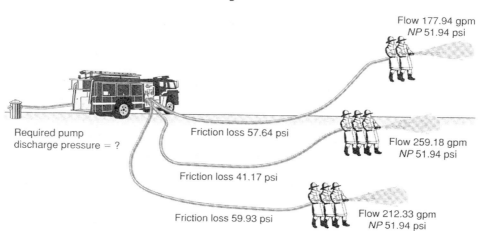

21. Add 2.34 + 34.72 + 6.345 + 113.45
22. Add 115.34 + 23.456 + 77.23 + 2.345
23. Add 94.56 + 2.345 + 774.34 + 45.724
24. Add 598.2 + 35.2 + 954.34 + 56.54
25. Add 55.66 + 34.78 + 83.45 + 66.21 + 43.24
26. Three lines have been taken off Engine 7 during a large fire. One line is a $2\frac{1}{2}$-inch with a tip supplying 210.57 gpm; the second is a 2-inch line supplying 195.45 gpm; and the third is a $1\frac{1}{2}$-inch line supplying 80.12 gpm. How much water is being supplied to the three lines?

    SOLUTION HELP—Add 210.57 + 195.45 + 80.12

27. On the first five days of May, Engine 9 traveled 8.4 miles on the first alarm, 3.4 on the second alarm, 5.2 on the third, and 4.6 on the fourth. How many miles did it travel on the four alarms?

    SOLUTION HELP—Add 8.4 + 3.4 + 5.2 + 4.6

28. During October, Engine 7 devoted three days to conducting fire prevention inspections. The first day they were out 2 hours and 10 minutes. The second day they were out 3 and a half hours, and on the third day they were out one hour and 45 minutes. How long did they spend on fire prevention inspections on the three days? Express the answer in decimals.

    SOLUTION HELP—Add 2 hours and 10 minutes + $3\frac{1}{2}$ hours + one hour and 45 minutes. Express the answer in decimals.

29. During the first four shifts of November, Engine 5 devoted 2.37 hours plus 3.15 hours, plus 2.19 hours, plus 2.56 hours to physical training. What was the total amount of time they devoted to physical training during these four shifts?

    SOLUTION HELP—Add 2.37 + 3.15 + 2.19 + 2.56

30. Engine 8 devotes the first portion of each shift to housework. The first five shifts of March they spent 3.2 plus 2.6 plus 4.5 plus 2.4 plus 3.1 to this activity. What was the total amount of time they devoted to housework during the first five shifts of March?

    SOLUTION HELP—Add 3.2 + 2.6 + 4.5 + 2.4 + 3.1

UNIT 3—SUBTRACTING DECIMALS

Principles of Subtracting Decimals

The only difference between subtracting whole numbers and subtracting decimals is that the decimals in the numbers should be lined up. It is also good practice to add zeros so that the minuend and the subtrahend have the same number of digits in the decimal. The proper procedure to follow is:

1. Write the figure to be subtracted vertically in the same manner as if subtracting whole numbers. Line up the decimal points.
2. If there are empty place-value positions, add zeros.
3. Complete the subtraction problem.

Problem 1: Subtract 454.67 from 856.721. Prove the answer by the conventional method of proving subtraction problems. Round the answer to the nearest hundredth.

Solution:

$$\begin{array}{r} 856.721 \\ -454.670 \\ \hline 402.051 \end{array}$$

Answer: rounded off = 402.05

In proving the answer, prove it prior to rounding off the answer, otherwise it will not check out.

The procedure for checking a subtraction problem was introduced in Chapter 4. It is repeated here as a review, using Problem 1 to demonstrate the procedure.

Problem 1:

$$\begin{array}{r} 856.721 \\ -454.670 \\ \hline 402.051 \end{array}$$

Step 1: Add the answer (difference) to the subtrahend.

$$402.051 + 454.67 = 856.721$$

Step 2: Check the addition against the minuend (856.721).

If both digits are the same, the problem has been solved correctly.

Problem 2: Subtract 45.673 from 102.3.

Solution:

$$\begin{array}{r} 102.300 \\ -45.673 \\ \hline 56.627 \end{array}$$

Check the answer.

Step 1: $56.627 + 45.673 = 102.3$

Step 2: Check the answer against the minuend (102.3).

If both the answer and the minuend are the same, the problem has been worked correctly.

Reader's Practice Problems—6.3

Work each of the following subtraction problems and check your answer using the conventional method.

1. 76.453
   − 43.7

2. 876.2
   − 445.78

3. 689.564
   − 67.22

4. 1,123.87
   −   99.98

The answers are in the Reader's Practice Problems section of the Appendix.

A PRACTICAL SOLUTION WHEN SUBTRACTING DECIMALS

When you subtract the amount you have written with a check from your checking account, in essence you are subtracting decimals to determine your new balance. There is a method that you should be aware of that can occasionally be used that simplifies the subtraction. Following is an example of this method.

You have a balance of $3,672.42 in your checking account. You then write a check for $39.96. Rather than complete the complicated subtraction, subtract $40 from the account. Because you have taken 4 cents too much from your account, add it back. The account would now total $3,632.46. Look for the opportunities to use this system. It makes your work much easier and less confusing. Once you start taking advantage of this method, you'll smile every time you use it.

UNIT 3—REVIEW QUESTIONS

1. What is the only difference between subtracting whole numbers and subtracting decimals?
2. What is the three-step procedure for solving subtraction decimal problems?
3. What procedure should be used for proving that a subtraction decimal problem has been worked correctly?

TEST 6.3

Solve the following problems and prove your answers. Round all answers to the nearest hundredth.

1. Subtract 43.961 from 157.72
2. Subtract 62.47 from 321.591
3. Subtract 145.62 from 576.59
4. Subtract 2,356.58 from 7,483.99
5. 123.56 − 34.98
6. 12.4853 − 9.4523
7. 254.2 − 76.568
8. 1.4 − .6742
9. 17.56 − 3.784
10. 1,256.78 − 32.089

Figure 6.4

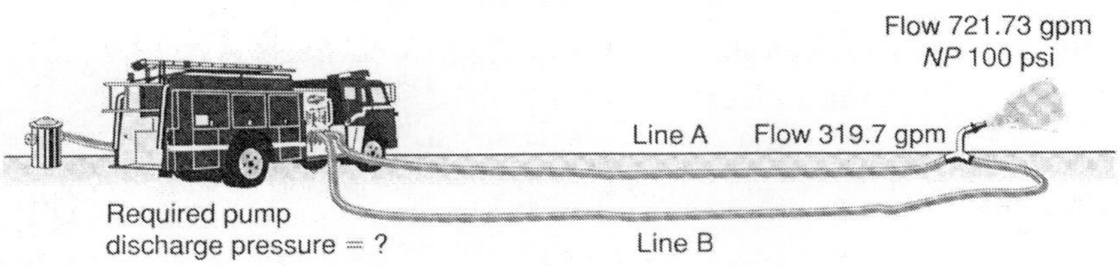

[Figure 6.4: Pumper with two lines. Line A Flow 319.7 gpm; total Flow 721.73 gpm, NP 100 psi at nozzle. Line B below. Required pump discharge pressure = ?]

11. You had $5461.17 in your checking account prior to writing a check for $99.99. What is the total you now have in your account?

    SOLUTION HELP—Subtract $100 from your checking account and add back .01.

12. You had $7,865.34 in your checking account prior to writing a check for $98.99. What is the total you now have in your account?

    SOLUTION HELP—Subtract $100 from the account and add back $1.01

13. An engine company in a large city had 5,463.7 miles on the apparatus odometer prior to receiving a move-up call. On arriving at the move-up location, the odometer indicated 5,481.6 miles. How far was it from one station to the other?

    SOLUTION HELP—Subtract 5,463.7 from 5,481.6

14. A busy truck company's odometer registered 12,834.6 miles at 8:00 a.m. on March 1. At 8:00 a.m. on April 1, the odometer read 13,425.8 miles. How many miles did the apparatus travel during the month of March?

    SOLUTION HELP—Subtract 12,834.6 from 13,425.8

15. A fire department's budget for the year was $6,889,465.78. Of this amount, $52,577.28 was spent for a new pumper. How much remained in the budget?

    SOLUTION HELP—Subtract $52,577.28 from $6,889,465.78

16. During the month of May, an engine company spent 78.6 hours in training. The following month they spent only 59.3 hours in training. How many more hours did they spend in training in May than in June?

    SOLUTION HELP—Subtract 59.3 from 78.6

17. In Figure 6.4, how much water is flowing through line B?

    SOLUTION HELP—Subtract 319.7 from 721.73

18. If line A in Figure 6.5 is shut down, what will be the flow from the pumper?

    SOLUTION HELP—Subtract 240.71 from 429.76

Figure 6.5

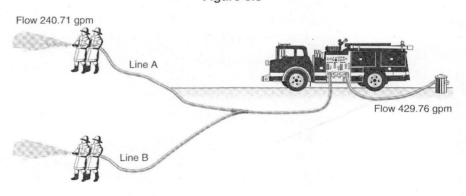

Figure 6.6

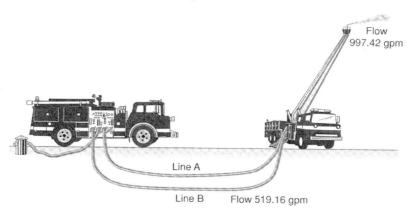

19. What is the flow through line A in Figure 6.6?
    SOLUTION HELP—Subtract 519.16 from 997.42
20. What is the flow from line C in Figure 6.7?
    SOLUTION HELP—Add 217.5 + 251.84 and subtract the answer from 682.06
21. 173.35 − 77.3
22. 256.34 − 17.5
23. 7823.45 − 573.23
24. 456.342 − 56.474
25. 65.432 − 55.463
26. On April 1, the odometer on Engine 7 read 13,005. On May 1, the odometer read 13,834.6. How many miles did the company run during the month of April?
    SOLUTION HELP—Subtract 13,005 from 13,834.6
27. Two 2-inch lines have been taken off Engine 5. Both lines are discharging 209.3 gpm. If the discharge on line 2 is reduced to 155.7 gpm, what will be the reduction in the pump's discharge?
    SOLUTION HELP—Subtract 155.7 from 209.3

Figure 6.7

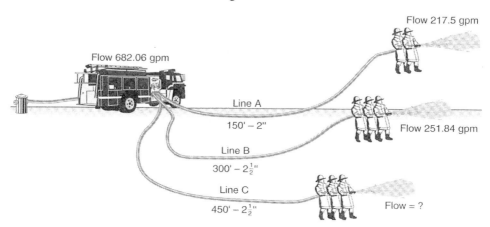

28. At 8:00 a.m. on March 9, the odometer on Engine 1 read 15,346.3. On the first of March, the odometer had read 15,188.7. How many miles had the company put on the apparatus during the first 8 days of March?

    SOLUTION HELP—Subtract 15,188.7 from 15,346.3

29. Firefighter Trlot set a goal of running an average of ten miles on each of his off-duty days. After his first five off-duty days, he had run 48.6 miles. How many more miles does he have to run to obtain his goal?

    SOLUTION HELP—Subtract 48.6 from 50

30. During the month of November, a young paramedic on Ambulance 3 worked 9.3 hours of overtime. During December, she worked only 6.4 overtime hours. How many more overtime hours did she work in November than in December?

    SOLUTION HELP—Subtract 6.4 from 9.3

UNIT 4—MULTIPLICATION OF DECIMALS

Principles of Multiplying Decimals

The process of multiplying decimals is the same as the process for multiplying whole numbers. The primary difference is locating the decimal points. The procedure to use is:

1. Complete the multiplication process as if all the digits are whole numbers.
2. Count the total number of decimal places in the factors that were multiplied.
3. Place the decimal point in the answer so that there are the same number of decimal places in the answer as there were found in Step 2.
4. Check the answer by the conventional method. Ignore the decimals during this process.
5. If possible, check the answer against an estimate of the answer. The estimate can be determined by rounding the numbers that were multiplied. For example, if the multiplication problem was to multiply 46.76 and 53.21, and you have an answer of 2,488.0996, that checks out to be correct. The only way your answer would be wrong is if the decimal point was incorrectly located. Round out the 46.76 to 50 and the 53.21 to 50. Check it by mentally multiplying 50×50, which equals 2,500. This is close to your answer. This indicates that your answer is correct. If your answer had been 248.8099 or perhaps 24,880.996, you would know that the decimal point was in the wrong location.

Problem: Multiply 33.63×36.42

Solution:

$$\begin{array}{r} 33.63 \\ \underline{36.42} \\ 6726 \\ 13452 \\ 20178 \\ \underline{10089} \\ 12248046 \end{array}$$

There were four decimal points in the multiplication problem. Count off four in the answer: 1224.8046

Check the answer.

$$
\begin{array}{r}
33.63 \\
\underline{36.42} \\
6726 \\
13452 \\
20178 \\
\underline{10089} \\
1224.8046
\end{array}
$$

Check the location of the decimal point by mentally multiplying:

$$30 \times 40 = 1200$$

Problem: Multiply $3.5 \times 6.2 \times 2.3$

Solution: First, multiply 35×62 (ignore the decimal point)

$$
\begin{array}{r}
35 \\
\underline{\times 62} \\
2170
\end{array}
$$

Then 21.70×2.3

$$
\begin{array}{r}
2170 \\
\underline{\times 23} \\
49910
\end{array}
$$

Now locate the decimal point. The numbers of the original problem have a total of three decimal points. Therefore, the decimal point will be located in the answer as 49.910.

Next, prove the decimal point is in the correct location by multiplying:

$$4 \times 6 \times 2 = 48$$

The answer is 49.910

Problem: Multiply 32.67×45.74. Round off your answer to the hundredths place. A word of caution: Check the answer prior to completing the rounding-off process.

Solution: 3267×4574 (ignore the decimal points) $= 14943258$

The total number of numbers past the decimal points in the problem is four.

Count off four in the answer: 1494.3258

Mentally check the correctness of the location of the decimal point by multiplying $30 \times 50 = 1,500$. It checks.

Reader's Practice Problems—6.4

Work the following decimal multiplication problems, and check each one. Also check the correction of the location of the decimal point. Round off each answer to the nearest hundredths place.

1. 67.5	2. 342.3	3. 36.67	4. 765.3	5. 452.7
$\times$ 42.4	$\times$ 7.8	$\times$ 89.63	$\times$ 2.83	$\times$ 4.71

The answers are in the Reader's Practice Problems section of the Appendix.

UNIT 4—REVIEW QUESTIONS

1. What is the primary difference between whole number multiplication problems and decimal multiplication problems?
2. What steps should be taken to solve a multiplication decimal problem?
3. What procedure should be used to prove that a decimal multiplication problem has been worked correctly?
4. What should be done with the decimal points when proving a decimal multiplication problem?
5. What is a good procedure to use to determine that the decimal point has been located correctly in the answer to a decimal multiplication problem?

TEST 6.4

1. Multiply 4.5 × 6.7
2. Multiply 23.4 × 3.9
3. Multiply 43.5 × 17.2
4. Multiply 156.23 × 21.3
5. Multiply 3.4 × 6.2 × 8.3

For problems 6–10, round all answers to the nearest thousandths place.

6. 23.76 × 14.31
7. 230.89 × 56.78
8. 7.98 × 3.65
9. 34.1 × 63.78 × 2.47
10. 10.1 × 3.99 × 34.2
11. There are 7.48 gallons of water in a cubic feet. To determine the number of gallons in a rectangular container, use the formula:

$$\text{Gallons} = 7.48\,V$$

where   V = volume in cubic feet

Problem: Determine the number of gallons of water in the container shown in Figure 6.8.
SOLUTION HELP—Multiply 7.48 × 4.2

12. There are 62.5 pounds of water in a cubic feet. To determine the number of pounds in a cylindrical container, use the formula:

$$\text{Pounds} = 62.5\,V$$

where   V = volume in cubic feet

Problem: What is the weight of the water in the container shown in Figure 6.9?
SOLUTION HELP—Multiply 62.5 × 3.72

13. To determine the pressure created by a column of water, use the formula:

$$P = .434\,H$$

where   H = head

The head is the vertical distance from the surface of the water to the point being considered.

Figure 6.8

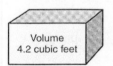

Figure 6.9

Figure 6.10

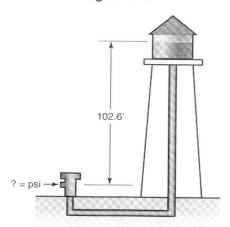

Problem: What is the pressure at the hydrant outlet shown in Figure 6.10? Round your answer to second decimal point.

SOLUTION HELP—Multiply .434 × 102.6

14. Using the information provided in Problem 13, what is the nozzle pressure on the nozzle being used in the three-story building shown in Figure 6.11? Round your answer to the second decimal point.

SOLUTION HELP—Multiply .434 × 119.3

Figure 6.11

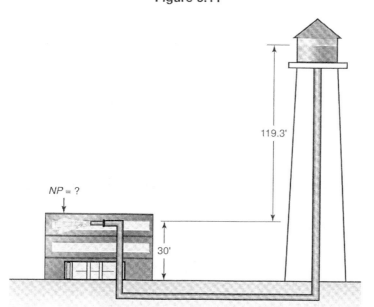

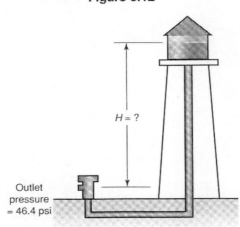

Figure 6.12

15. To determine the head when the pressure at an outlet is known, use the formula:

$$H = 2.304\, P$$

where P = pressure at the outlet

Problem: What is the vertical distance from the hydraulic outlet shown in Figure 6.12 and the surface of the water in the tank? Round your answer to the second decimal point.
SOLUTION HELP—2.304 × 46.4

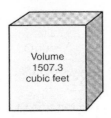

Figure 6.13

16. The number of gallons of water that a rectangular container can hold is determined by the formula:

$$\text{Gallons} = 7.48\, V$$

where V = volume in cubic feet

Problem: How many gallons of water can the container shown in Figure 6.13 hold?
SOLUTION HELP—Multiply 7.48 × 1507.3

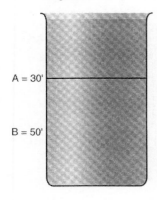

Figure 6.14

17. Fluid pressure of a liquid in a container acts perpendicular to the surface on which it acts. The pressure is equal to .434 H. H refers to the head, which is defined as the vertical distance from the surface of the water to the point upon which it acts. What is the pressure at point A and at point B in Figure 6.14?

SOLUTION HELP—Multiply .434 × 30 and .434 × 50

18. When a line taken off a pumper is being used at a higher elevation than the pumper, a pressure is exerted back against the pumper. This pressure is called back pressure and is equal to .434 H. When a line taken off a pumper is being used at an elevation below the pumper, the pressure created by the head assists the pumper. This pressure is referred to as forward pressure and is equal to .434 H. What is the forward pressure at the nozzle shown in Figure 6.15?

SOULTION HELP—Multiply .434 × 130

19. The volume of a rectangular container can be determined by the formula:

$$\text{Volume} = lwh$$

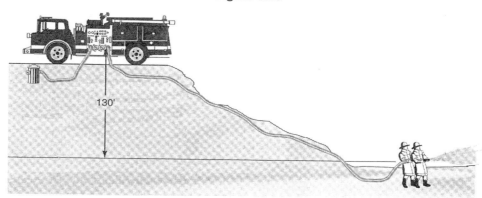

Figure 6.15

where   l = length
         w = width
         h = height

What is the volume of the rectangular container shown in Figure 6.16? Round your answer to the second decimal point.

SOLUTION HELP—Multiply 14.3 × 10.2 × 8.1

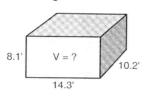

Figure 6.16

20. The intake gauge on the pumper shown in Figure 6.17 reads 14 inches of mercury. The lift can be determined by the formula:

$$H = 1.13 \, Hg$$

where   H = height (lift in feet)

         Hg = inches of mercury (on intake gauge)

What is the lift?

SOLUTION HELP—Multiply 1.13 × 14

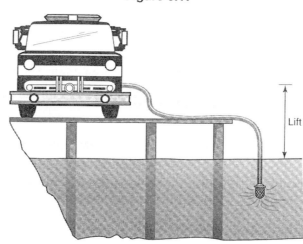

Figure 6.17

21. 34.4 × 67.2
22. 17.9 × 83.2
23. 64.2 × 56.4
24. 17.5 × 82.3
25. 112.7 × 634.5
26. To determine the number of gallons of water that a container can hold, use the formula:

$$\text{Gallons} = 7.48\, V$$

where   V = container capacity

How many gallons can a rectangular container that measures 10 feet by 8 feet by 25 feet hold? The volume of a rectangular container can be obtained by the formula: volume = length × width × height.

SOLUTION HELP—Multiply 7.48 × 10 × 8 × 25

27. To obtain the number of pounds of water that a rectangular container can hold, use the formula:

$$\text{weight} = 62.5\, V$$

where   V = volume in cubic feet

How much weight will be in a rectangular container that has a volume of 2,000 cubic feet?

SOLUTION HELP—Multiply 62.5 × 2000

28. To obtain the pressure created by a column of water, use the formula:

$$\text{Pressure} = .434\, H$$

where   H = head

What would be the static pressure on a hydrant that is located 103 feet below the level of the water supplying the hydrant?

SOLUTION HELP—Multiply .434 × 103

29. To determine the head when the pressure at an outlet is known, use the formula:

$$H = 2.304\, P$$

where   P = pressure at the outlet

What would be the head supplying a hydrant that has a static pressure of 54 psi?

SOLUTION HELP—Multiply 2.304 × 54

30. The discharge from a nozzle tip can be determined by the formula:

$$\text{Discharge} = 29.7\, D^2 \sqrt{P}$$

where   D = diameter of the tip
        P = nozzle pressure

What would be the discharge from a 1.25-inch tip that has a nozzle pressure of 50 psi? (the square root of 50 = 7.07)

SOLUTION HELP—Multiply 29.7 × 1.25 × 1.25 × 7.07

UNIT 5—DIVISION OF DECIMALS

REVIEW WORDS

dividend—The number to be divided.

divisor—The number that will be used to divide the dividend into equal parts.

quotient—The answer to the division problem.

The Principle of Dividing Decimals

The setup for a division problem can be illustrated as follows:

$$\text{divisor} \quad 3\overline{)9} \quad \begin{array}{c} 3 \\ \end{array} \quad \text{quotient} \\ \text{dividend}$$

As with multiplication of decimal problems, the primary difference between working with a division problem involving whole numbers and a division problem having decimals is where the decimals are located. However, the procedure is relatively simple.

Both the divisor and the dividend in the following example have decimals in their makeup.

$$2.36\overline{)23.376}$$

Start with the divisor. Move the decimal point to the right until the divisor becomes a whole number.

$$236\overline{)23.376}$$

It took two moves to complete it. The next step is to move the decimal in the dividend the same number of spaces (two).

$$236\overline{)2337.6}$$

Now place a decimal in the quotient directly above the decimal in the dividend.

$$236\overline{)2337.6}$$

The problem is now set up. Proceed as if there were no decimals in the problem.

A different look: The following example has a decimal in the divisor; however, the dividend is a whole number.

$$4.39\overline{)725}$$

The first step is the same. Move the decimal point in the divisor to the right until the divisor becomes a whole number.

$$439\overline{)725}$$

It took two moves to complete it. The next step is to add as many zeros to the dividend as the decimal in the divisor was moved (two), and add a decimal point after the last zero.

$$439\overline{)72500.}$$

If the problem is to be solved correctly to the hundredths place, then add at least two more zeros past the decimal in the dividend, and place a decimal in the quotient directly above the decimal in the dividend.

$$439\overline{)72500.00}$$

The problem is now set up. Proceed as if there were no decimals in the problem.

Problem: 3.2)̄45.7

Set up the problem: 32)̄457.00

Solve the problem

```
            14.28
        32)457.00
            32
            ---
            137
            128
            ---
             90
             64
            ---
            260
            256
            ---
              4   remainder
```

Check the problem prior to eliminating the 4.

```
            14.28
        32)457.00
            32
            ---
            137
            128
            ---
             90
             64
            ---
            260
            256
            ---
              4
```

To check the problem:

First, multiply the divisor × the quotient: 32 × 14.28 = 456.96

Second, add the remainder (.04) to the result of the multiplication.

$$456.96 + .04 = 457$$

If this result and the original problem are the same, the problem has been worked correctly.

Reader's Practice Problems—6.5

Solve the following decimal division problems. Provide an answer for each of them to the second decimal point. Prove that your answer is correct.

1. Divide 76.454 by 2.61
2. Divide 3.478 by 12.95
3. 46.739 ÷ .23
4. 634)̄5459.22315
5. 2.326)̄8764.678

The answers are in the Reader's Practice Problems section of the Appendix.

UNIT 5—REVIEW QUESTIONS

1. What are the two primary differences between working with division problems involving whole numbers and division problems involving decimals?
2. Explain the procedure for dividing decimals.
3. Explain how to prove a division problem involving decimals by the "excess of nines" method.

TEST 6.5

Work all problems to the hundredths place. Prove that your answer is correct.

1. $4.3\overline{)67.43}$
2. $12.3\overline{)4.72}$
3. $67.7\overline{)126.784}$
4. $2.7\overline{)5.546}$
5. $436\overline{)549.72}$
6. Divide 423.78 by 32.45
7. Divide 34.56 by 2.38
8. $98 \div 3.5$
9. $16 \div 1.9$
10. $3457.23 \div 39.2$

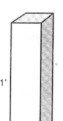

Figure 6.18

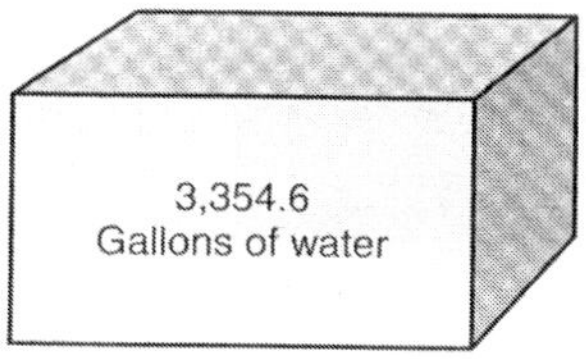

3,354.6 Gallons of water

11. There are 7.48 gallons of water in a cubic feet. To determine the number of cubic feet in a rectangular container when the number of gallons of water in the container is known, use the formula:

$$\text{Cubic feet} = \frac{\text{gallons}}{7.48}$$

Problem: How many cubic feet are there in the rectangular container shown in Figure 6.18?

SOLUTION HELP—Divide 3,354.6 by 7.48

12. A rectangular container measuring 1 inch by 1 inch at its base and 1 feet high will hold .434 pounds of water. A 1-cubic feet rectangular container holds 62.5 pounds of water. Using the information in Figure 6.19 as a guide, how many of the containers shown in Figure A will there be in Figure B?

SOUTION HELP—Divide 62.5 by .434

13. To determine the pressure at a given point, use the formula:

$$P = .434 \, H$$

where   H = head

Figure 6.19

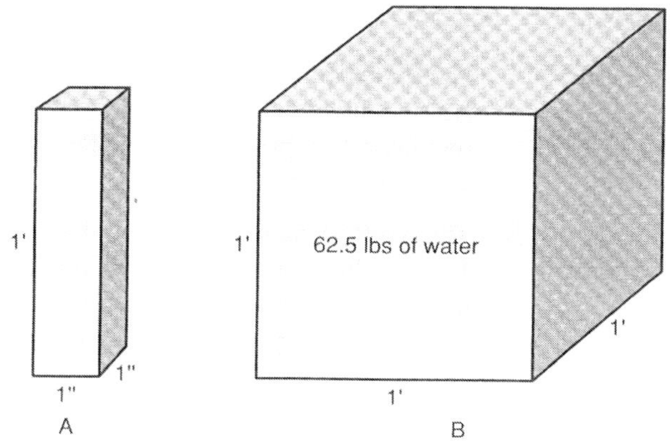

A

B — 62.5 lbs of water

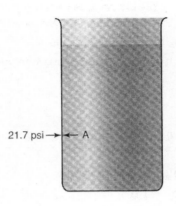

Figure 6.20

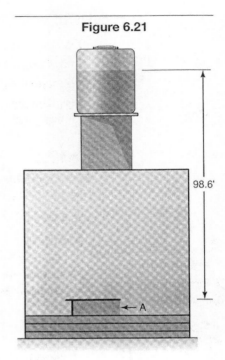

Figure 6.21

Head is the vertical distance from the surface of the water in a container to the point being considered. The pressure at point A in Figure 6.20 is 21.7 psi. What is the distance from the surface of the water to point A?

SOLUTION HELP—Divide 21.7 by .434

14. To determine the pressure at an outlet, use the formula:

$$P = .434\,H \quad \text{or} \quad P = H/2.304$$

Problem: Use the information in Figure 6.21 to determine the pressure at point A.

SOLUTION HELP—Multiply .434 × 98.6

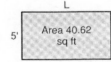

Figure 6.22

15. The formula for determining the area of a rectangle is:

$$\text{area} = lw$$

where the area is in square feet

l = length

w = width

If the area and the width are known and it is desired to determine the length, the length can be determined by the formula $l = \text{area}/w$. What is the length of the rectangle shown in Figure 6.22?

SOLUTION HELP—Divide 40.62 by 5

16. There are 31 days in January. If every month had the same number of days as January, how many months would there be in a year counting January? For practice, provide an answer to the second decimal point.

SOLUTION HELP—Divide 365 by 31

17. A paramedic on an ambulance that runs out of Engine 6's station works 8.2 hours on Monday, 8 hours on Tuesday, 8.4 hours on Wednesday, 8 hours on

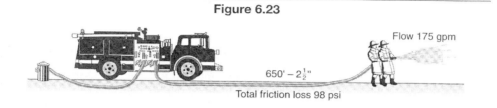

Figure 6.23

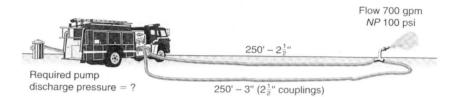

Figure 6.24

Thursday, 8.9 hours on Friday, and 9.2 hours on Saturday. What was the average number of hours she worked each day?

SOLUTION HELP—Add 8.2 + 8 + 8.4 + 8 + 8.9 + 9.2. Then divide the answer by 6. This provides the average hours worked for the 6 days.

18. How much is the friction loss for each 100 feet of hose shown in Figure 6.23?

SOLUTION HELP—Divide 98 by 6.50

19. During a warehouse fire, the portable monitor shown in Figure 6.24 discharged a total of 22,300 gallons of water. How long was the portable monitor used on the fire?

SOLUTION HELP—Divide 22,300 by 700

20. If it were possible to use all of the water shown in the swimming pool in Figure 6.25, how long would it take to empty the pool?

SOLUTION HELP—Divide 10,000 by 165

21. $34 \div 32.5$
22. $78 \div 44.7$
23. $23.7 \div 25.6$
24. $16.4 \overline{)44.35}$
25. $57.4 \overline{)124.5}$

26. Engine 3 used a portable monitor on a furniture store fire. The monitor was equipped with a $1\frac{3}{4}$-inch tip discharging 800 gpm. During the fire, the monitor discharged 6,853.76 gallons of water. How long was the monitor used on the fire?

SOLUTION HELP—Divide 6,853.76 by 800

Figure 6.25

27. How long would it have taken three 2 1.2-inch lines equipped with $1\frac{1}{4}$-inch tips, each discharging 325.7 gpm, to discharge the same amount of water as the portable monitor?

    SOLUTION HELP—Divide 6853.76 by 977.1 (3 × 325.7 = 977.1)

28. The RPDP for a pumper supplying a $1\frac{1}{8}$-inch tip with a nozzle pressure of 45 psi can be determined by the formula RPDP = FL + NP. Engine 5 requires a discharge pressure of 143.4 in order to supply the $1\frac{1}{8}$ tip through 450 feet of $2\frac{1}{2}$-inch hose. What is the friction loss in each section of $2\frac{1}{2}$-inch hose? (Remember that the FL stands for friction loss in the line and that each section of hose is 50 feet in length.)

    SOLUTION HELP—Divide 98.4 (the friction loss in the hose) by 9 (the number of sections of hose used)

29. How was the friction loss in the hose determined in Question 28?

    SOLUTION HELP—By subtracting the nozzle pressure (45) from the discharge pressure (143.4) 143.4 − 45 = 98.4

30. A swimming pool holds 12,543.7 gallons of water. By using siphon ejectors, Engine 7 is able to discharge 978 gallons per hour from the pool. How long will it take to empty the pool?

    SOLUTION HELP—Divide 12,543.7 by 978

Chapter Summary

The first portion of this chapter explained the procedures for working with decimals. The key words were defined and the rounding of numbers introduced. Instructions were provided on how to convert a fraction to a decimal and to convert a decimal to a fraction.

The chapter was then divided into four separate units: Unit 2 described the procedure for the addition of decimals, Unit 3 the procedure for the subtraction of decimals, Unit 4 the procedure for multiplying decimals, and Unit 5 the procedure for dividing decimals. All five units also had problems for the reader to solve.

CHAPTER 7

SOLVING COMBINATION PROBLEMS

KEY TERMS

braces
brackets
exponent
parentheses

OBJECTIVES AND GOALS

Upon completing this chapter, the reader should be able to solve problems involving more than a single type of operation.

INTRODUCTION

You have been exposed to the processes of adding, subtracting, multiplying, dividing, and determining the square root of whole numbers. You have also been exposed to adding, subtracting, multiplying, and dividing fractions and decimals. With the ability to determine the square root of whole numbers, it is assumed that you are capable of working individual problems in all of the other areas. However, mathematical problems are not always limited to a single type of operation. Some problems require combining two or more operations to determine the answer. Unless you are familiar with the proper sequence of operations, you may have trouble with some of these problems. This chapter helps you in working with combination problems.

For example, examine the following problem:

$$3 + 6 \times 3\frac{1}{2} - \frac{2}{3}$$

Suppose you add $3 + 6 = 9$. Then you multiply $9 \times 3\frac{1}{2} = 31\frac{1}{2}$. Then subtract $31\frac{1}{2} - \frac{2}{3} = 30\frac{5}{6}$.

Or you might add $3 + 6 = 9$. Then $3\frac{1}{2} - \frac{2}{3} = 2\frac{5}{6}$. Then $9 \times 2\frac{5}{6} = 25\frac{1}{2}$.

Or you may decide to multiply $6 \times 3\frac{1}{2} = 21$. Then $21 - \frac{2}{3} = 20\frac{1}{3}$. Then $20\frac{1}{3} + 3 = 23\frac{1}{3}$.

Do you have a choice in determining the correct answer to the problem? Of course, the answer is "no." The problem has been set up to work in the proper order. If it is worked in any other order, the answer will be wrong.

Basic Sequence for Working Combination Problems

First, it is necessary to understand the meaning of three symbols used with mathematical problems. These symbols are referred to as grouping symbols, and assist in indicating the order of operations:

parentheses = ( )
brackets = [ ]
braces = { }   The use of braces is limited.

RULES OF THE GAME

The rules of the game for solving combination mathematical problems are in reality a pecking order of who goes first. The rules are established in a set of principles referred to as the Order of Operations. If an individual working a combination problem does not comply with this pecking order, it is most likely that he or she will arrive at the wrong answer to the problem. It is important that the reader understand and work within the established Order of Operations.

ORDER OF OPERATIONS

The correct order of operations is a four-step process. Each step should be worked from left to right.

Rule 1: First, do all operations inside parentheses or other grouping symbols. The operations within the parentheses or other grouping symbols should comply with the Order of Operations.

Rule 2: Next, complete any expressions with exponents or roots.

Rule 3: Multiply and divide, proceeding from left to right.

Rule 4: Add and subtract, proceeding from left to right.

Use the following phrase to remember the proper order of operations.

PLEASE	parentheses and brackets
EXPECT	exponents and square roots
MY	multiply and divide
AUNT	add and subtract

Another important point to remember is that exponents and roots have equal value. One is not always worked prior to the other. The one that is encountered first when operating from left to right has precedence. The same concept applies to multiplying and dividing, and with adding and subtracting.

WORKING WITH RULE 1: PARENTHESES AND OTHER GROUPING SYMBOLS

Following is the correct order for completing operations inside parentheses or other grouping symbols.

1. Complete the operations within the parentheses first using the priority system outlined above.
2. When the operation is complete, replace the parentheses with the result.
3. Complete the operations within the brackets using the priority system previously outlined.
4. Replace the brackets with the result.

   For purposes of illustration, a sample problem using all three symbols is used.

Problem: $\{8 - [2 \times (5 - 2)]\} =$

Solution: $\{8 - [2 \times (5 - 2)]\}$

> *Step 1.* Complete the operation within the parentheses (subtract $5 - 2 = 3$).
>
> *Step 2.* Replace the parentheses with the 3. The problem now appears as:
> $$\{8 - [2 \times 3]\}$$
>
> *Step 3.* Complete the operation within the inner brackets.
> $$2 \times 3 = 6$$
>
> *Step 4.* Replace the brackets with the 6 and remove the braces. The problem now appears as:
> $$8 - 6 =$$
>
> *Step 5.* Solve the problem.
> $$8 - 6 = 2$$

Problem: $5 + (6 - 2) - 3 =$

Solution:

> *Step 1.* Complete the operation within the parentheses (Rule 1) $(6 - 2) = 4$.
>
> *Step 2.* Replace the parentheses with the 4. The problem is now ready to be solved.
> $$5 + 4 - 3 = 6$$

Problem: $17 - (3 \times 4) + (8 \div 2) - 5 =$

Solution:

> *Step 1.* Complete the operation within the first parentheses (Rule 1) $(3 \times 4) = 12$.
>
> *Step 2.* Replace the first parentheses with the 12. The problem now appears as:
> $$17 - 12 + (8 \div 2) - 5 = 4$$
>
> *Step 3.* Complete the operation within the second parentheses (Rule 1).
> $$(8 \div 2) = 4$$
>
> *Step 4.* Replace the second parentheses with the 4. The problem is now ready to be solved.
> $$17 - 12 + 4 - 5 = 4$$
>
> *Step 5.* $\qquad\qquad\quad 17 - 12 + 4 - 5 = 4$
> Then $\qquad\qquad\qquad\qquad\qquad\quad 4 = 4$
> Subtract 4 from each side $\qquad 4 - 4 = 4 - 4$
> $\qquad\qquad\qquad\qquad\qquad\qquad\quad 0 = 0$

WORKING WITH RULE 2: EXPONENTS AND ROOTS

An *exponent* is a short method for indicating how many times a number is to be multiplied by itself. The root that is found most often in fire department hydraulics is the square root.

Problem: $4 + 3 + 2^2 =$

Solution:

> *Step 1.* Square the 2 (Rule 2) $(2^2 = 4)$.
>
> *Step 2.* Replace the 2 squared with the 4. The problem is now ready to be solved.
> $$4 + 3 + 4 = 11$$

Problem: $5 - 2 + \sqrt{16}$

Solution:

Step 1. Determine the square root of 16 (Rule 2) ($\sqrt{16} = 4$).

Step 2. Replace the $\sqrt{16}$ with the 4. The problem is now ready to be solved.

$$5 - 2 + 4 = 7$$

Problem: $4^2 + 2\sqrt{9} - 3^2 + 5 =$

Note: $2\sqrt{9}$ is a shortcut method of writing $2 \times \sqrt{9}$.

Solution:

Step 1. Determine 4^2 (Rule 2) ($4^2 = 16$).

Step 2. Replace the 4^2 with 16. The problem now appears as:

$$16 + 2\sqrt{9} - 3^2 + 5 =$$

Step 3. Determine the $\sqrt{9}$ (Rule 2): $\sqrt{9} = 3$.

Step 4. Replace the $\sqrt{9}$ with the 3. The problem now appears as:

$$16 + 2 \times 3 - 3^2 + 5 =$$

Step 4. Determine 3^2 (Rule 2): $3^2 = 9$.

Step 5. Replace the 3^2 with the 9. The problem now appears as:

$$16 + 2 \times 3 - 9 + 5 =$$

Step 6. Complete the multiplication of 2×3 (Rule 3): $2 \times 3 = 6$. Replace the 2×3 with a 6. The problem is now ready to be solved.

$$16 + 6 - 9 + 5 = 18$$

WORKING WITH RULE 3: MULTIPLICATION AND DIVISION

Problem: $6 - 4 + 3 \times 6$

Solution:

Step 1. Complete the multiplication of 3×6 (Rule 3): $3 \times 6 = 18$.

Step 2. Replace the multiplication with the 18. The problem is now ready to be solved.

$$6 - 4 + 18 = 20$$

Problem: $8 + 2 - 6 \div 2$.

Solution:

Step 1. Complete the division portion of the problem (Rule 3): $6 \div 2 = 3$.

Step 2. Replace the division portion with the 3. The problem is now ready to be solved.

$$8 + 2 - 3 = 7$$

Problem: $5 \times 6 - 18 \div 3$

Solution:

Step 1. Complete the multiplication portion of the problem

$$5 \times 6 = 30$$

Step 2. Replace the 5×6 with a 30.

$$30 - 18 \div 3$$

Step 3. Complete the division portion of the problem.

$$18 \div 3 = 6$$

Step 4. Replace the $18 \div 3$ with a 6.

Step 5. The problem is now ready to be solved.

$$30 - 6 = 24$$

Problem: $3^2 + (5 - 4 + 7) - \sqrt{16} - 24 \div 3$

Step 1. Work the problem within the parentheses.

$$(5 - 4 + 7) = 8$$

Step 2. Replace the parentheses with an 8.

$$3^2 + 8 - \sqrt{16} - 24 \div 3$$

Step 3. Square the 3 (Rule 2): $3^2 = 9$.

Step 4. Replace the 3^2 with the 9.

$$9 + 8 - \sqrt{16} - 24 \div 3$$

Step 5. Solve the $\sqrt{\ }$ of 16 (Rule 2): $\sqrt{16} = 4$

Step 6. Replace the $\sqrt{16}$ with a 4.

$$9 + 8 - 4 - 24 \div 3$$

Step 7. Divide $24 \div 3$ (Rule 3): $24 \div 3 = 8$

Step 8. Replace the $24 \div 3$ with the 8.

Step 9. The problem is now ready to solve.

$$9 + 8 - 4 - 8 = 5$$

Problem: $24 - (2 + 3 \times 4) + 6 + \sqrt{9} \times (7 - 4) + 2 \times 4^2 - 5 =$

Step 1. Complete the operations within the parentheses working from left to right.

$$\text{Rule 1: } (2 + 3 \times 4)$$

First, multiply $3 \times 4 = 12$

Next, replace the multiplication inside the parentheses with the 12. The problem inside the parenthesis now appears as:

$$(2 + 12) \quad 2 + 12 = 14$$

Step 2. Replace the first parentheses in the problem with the 14. The problem now appears as:

$$24 - 14 + 6 + \sqrt{9} \times (7 - 4) + 2 \times 4^2 - 5 =$$

Step 3. Complete the operation within the second parentheses.

$$(7 - 4) = 3$$

Step 4. Replace the second parentheses with the 3. The problem now appears as:

$$24 - 14 + 6 + \sqrt{9} \times 3 + 2 \times 4^2 - 5 =$$

Step 5. Solve $\sqrt{9}$

$$\sqrt{9} = 3$$

Step 6. Replace the $\sqrt{9}$ with the 3. The problem now appears as:

$$24 - 14 + 6 + 3 \times 3 + 2 \times 4^2 - 5 =$$

Step 7. Solve 4^2 (Rule 2): $4^2 = 16$.

Step 8. Replace the 4^2 with the 16. The problem now appears as:

$$24 - 14 + 6 + 3 \times 3 + 2 \times 16 - 5 =$$

Step 9. Multiply 3×3

$$3 \times 3 = 9$$

Step 10. Replace the 3×3 with the 9. The problem now appears as:

$$24 - 14 + 6 + 9 + 2 \times 16 - 5 =$$

Step 11. Multiply 2×16

$$2 \times 16 = 32$$

Step 12. Replace the 2×16 with the 32. The problem now appears as:

$$24 - 14 + 6 + 9 + 32 - 5$$

Step 13. The problem is now ready to be solved.

$$24 - 14 + 6 + 9 + 32 - 5 = 52$$

WORKING WITH RULE 4: ADDITION AND SUBTRACTION

We worked with Rule 4 as the last step of the first three rules.

Reader's Practice Problems—7.1

Work the following problems. Follow the rules of the game when solving the problems.

1. $8 - 4 + (6 \times 5) =$
2. $7 - [3 + 4 - (7 \times 2) + 10] =$
3. $(7 \times 7) + 3 \times 3 + (8 - 4) =$
4. $9 + 5^2 - 6 \times 7 =$
5. $9 - 3 + \sqrt{25} + 6 =$
6. $3 \times \sqrt{4} + 21 + 3 \times 3^2 =$
7. $6 \times 2 + 8 - 21 \div 3 =$
8. $14 \div 7 + 23 - 5 + 6 \times 6$
9. $3 + (2 \times \sqrt{16}) - 5 \times 5 + 8^2 =$
10. $[2 \times 4 + (25 \div 5) - (\sqrt{16} \times 3^2)] =$

The answers are found in the Reader's Practice Problems section of the Appendix.

REVIEW QUESTIONS

1. What is this? ( )
2. What is this? [ ]

3. What is this? { }
4. Which operation on a combination problem should be completed first?
5. Which operation on a combination problem should be completed second?
6. Which operation on a combination problem should be completed third?
7. What is an exponent?
8. When working the problem $6 - 4 + \sqrt{25}$, what operation should be completed first?
9. When working the problem $8 - 6 + 3 \times 7$, what operation should be completed first?
10. When working the problem, $9 + 4 - 8 \div 2$, what operation should be completed first?

Test 7

1. $3 + 5 \times 2$
2. $3^2 \times 3 - 2$
3. $\sqrt{16} + 3 + 6$
4. $28 - 4 \times (7 - 2)$
5. $\frac{1}{2} + \frac{1}{3} \times \frac{3}{5}$
6. $4^2 \times \frac{2}{3} \times \frac{1}{2}$
7. $(\frac{1}{2} \times \frac{2}{3}) + 4 - \frac{1}{2} + (\frac{2}{9} \times \frac{3}{4})$
8. $\sqrt{4} \times 2 + 3^2 + 2 \times (\frac{2}{3} \times \frac{1}{3})$
9. $4 \times [6 + 2 \times (1\frac{3}{4} \times \frac{2}{3})]$
10. $\frac{2}{3} \times 6\frac{1}{2} \times \frac{4}{5} + \frac{3}{4} \times (\frac{1}{2} \times \sqrt{4})$

Important Information

Following is a summary of numbers that will be useful in solving several of the practical problems that follow. It is suggested that firefighters and future firefighters memorize these numbers. Emergency Medical Personnel should use them for reference.

- Water weighs approximately 62.5 pounds per cubic feet.
- There are approximately 7.48 gallons in a cubic feet.
- One gallon of water weighs approximately 8.35 pounds.
- There are 1,728 cubic inches in a cubic feet.
- One gallon of water contains approximately 231 cubic inches.
- A column of water measuring one inch by one inch at its base and one feet high weighs approximately .434 pounds.
- A column of water measuring one inch by one inch at its base and is 2.304 feet high weighs approximately one pound.

The next word problems use actual hydraulic formulas. These hydraulic word problems are presented in such a way that the solver starts with a hydraulic formula known to him or her, and then selects the information that is needed to solve the formula. Because all of the formulas may not be known to the reader, they are given when necessary. The accompanying figures provide the information needed to solve the formulas. Each problem is to be solved according to its priority in working combination problems.

Figure 7.1

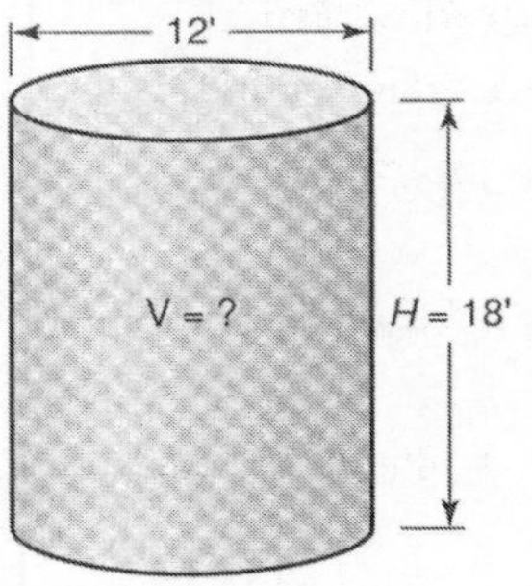

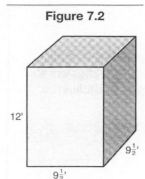

11. This problem combines the operations of the multiplication of decimals and whole numbers and determining the square of a number.

    The volume of a cylindrical container can be determined by the formula:

    $$\text{Volume} = .7854\, d^2 h$$

    where d = diameter
           h = height

    What is the volume of the container shown in Figure 7.1?

    SOLUTION HELP— From Figure 7.1, the diameter is 12 feet and the height is 18 feet.

    Place these into the formula.

    $$V = .7854 \times 12 \times 12 \times 18$$

    Solve the problem.

    $$V = .7854 \times 12 \times 12 \times 18$$
    $$= 2035.76 \text{ cubic feet}$$

12. This problem combines the multiplication of fractions with the multiplication of whole numbers, and the multiplication of decimals with whole numbers.

    The volume of a rectangular container can be solved by the formula:

    $$V = lwh$$

    where l = length
           w = width
           h = height

    The number of gallons in a cubic feet is 7.48.

    How many gallons of water will the rectangular container shown in Figure 7.2 hold?

    SOLUTION HELP—From Figure 7.2, the length is $9\frac{1}{2}$ feet, the width is $9\frac{1}{2}$ feet, and the height is 12 feet.

    Place these into the formula.

    $$V = 9.5 \times 9.5 \times 12$$

    Solve the problem.

    $$V = 9.5 \times 9.5 \times 12$$
    $$= 1{,}083 \text{ cubic feet}$$

    Because there are 7.48 gallons in a cubic feet, multiply 7.48×1083.

    $$7.48 \times 1{,}083 = 8{,}100.84 \text{ gallons}$$

Figure 7.2

13. This problem combines the operations of the multiplication of decimals, the squaring of a number, the multiplication of a whole number, and division.

    How many gallons of water will the hose line shown in Figure 7.3 hold?
    Hint: This is a cylindrical container. You are working in inches, not feet,

Figure 7.3

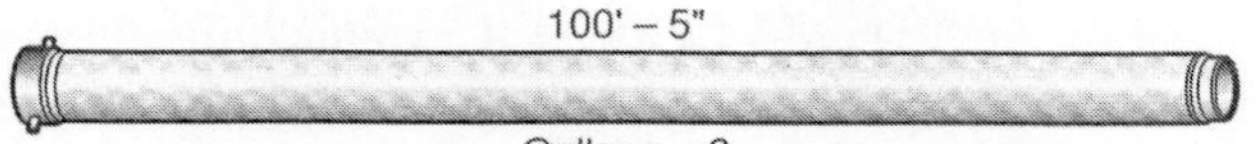

100' – 5"

Gallons = ?

including the length of the hose line. The formula to determine the gallon capacity of a hose line is:

$$\text{Gallon capacity} = \frac{\text{volume (cubic inches)}}{231}$$

SOLUTION HELP—First, determine the volume of the hose in cubic inches.

The hose is a cylindrical container that has a diameter of 5 inches and is 100×12 inches in length. The formula for the volume is:

$$V = .7854 \times d^2 \times \text{length (see Problem 11)}$$

Place the measurements into the formula.

$$V = .7854 \times 5^2 \times (100 \times 12)$$

Solve the formula.

$$V = (.7854)(5)(5)(1200)$$

First, square the 5.

$$5^2 = 25$$

Next, place the 25 into the formula.

$$V = (.7854)(25)(1200)$$

Solve the formula.

$$V = 23{,}562$$

Place the answer in the gallons capacity formula.

$$\text{Gallons capacity} = \frac{\text{volume}}{231}$$

$$\text{Solve the formula's gallons capacity} = \frac{23{,}562}{231} = 102 \text{ gallons}$$

14. This problem involves the combined operations of squaring a number, adding whole numbers, and determining the square root of a number.

There are three separate formulas for determining a 90° right triangle (see Figure 7.4). Occasionally, two sides of a right triangle are known and it is desired to determine the length of the third side. In such a situation, one of the following formulas will be used.

$$C = \sqrt{A^2 + B^2}$$
$$A = \sqrt{C^2 - B^2}$$
$$B = \sqrt{C^2 - A^2}$$

Using the appropriate formula, what will C equal if A = 4 feet and B = 5 feet?
SOLUTION HELP—Place the known sides into the formula

$$C = \sqrt{A^2 + B^2}$$
$$= \sqrt{4^2 + 5^2}$$

Figure 7.4

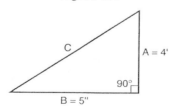

First, square the 4 and then the 5.

$$4^2 = 16$$
$$5^2 = 25$$

Place the results into the formula.

$$C = \sqrt{16 + 25}$$
$$C = \sqrt{41}$$

Solve the formula.

$$C = 6.40 \text{ feet}$$

15. This problem combines the operations of determining the square root of a number and the multiplication of decimals.

    Velocity flow refers to how fast water is moving through a hose or out a nozzle or other opening. The formula for determining the velocity flow is:

    $$V = 12.14\sqrt{P}$$

    where   V = velocity flow in feet per second
             P = nozzle pressure

    What is the velocity flow from the tip being used in Figure 7.5?
    SOLUTION HELP—The nozzle pressure from the figure is 50 psi.

    First, place the nozzle pressure into the formula.

    $$V = 12.14\sqrt{50}$$

    Next, determine the square root of 50.
    The square root of 50 is 7.07 (from Unit 3 in Chapter 4).
    Place the 7.07 into the formula to replace the $\sqrt{}$.

    $$V = (12.14)(7.07)$$

    Solve the formula.

    $$V = (12.14)(7.07)$$
    $$= 85.83 \text{ fps}$$

16. This problem combines the operations of squaring a number, converting a fraction to a decimal, and multiplying both a decimal and a whole number.

    For every action there is an equal and opposite reaction. A certain amount of force is exerted when water is discharged from a tip. The opposite force reacting against the firefighters holding the nozzle is called the nozzle reaction. The formula for determining the nozzle reaction is:

    $$NR = 1.57 \, D^2 P \quad \text{(answer in pounds)}$$

Figure 7.5

Figure 7.6

where   D = diameter of the nozzle tip

P = nozzle pressure

What is the nozzle reaction of the situation shown in Figure 7.6?

SOLUTION HELP—From Figure 7.6, the nozzle tip is $\frac{5}{8}$ inch, and the nozzle pressure is 45 psi.

First, place the nozzle tip size and the nozzle pressure into the formula.

$$NR = (1.57)\left(\frac{5}{8}\right)^2(45)$$

$$= \frac{(1.57)(5)(5)(45)}{1 \quad 8 \quad 8 \quad 1}$$

Next, solve the formula.

$$NR = \frac{(1.57)(5)(5)(45)}{(8)(8)}$$

$$= \frac{1766.25}{64}$$

$$= 27.6 \text{ pounds}$$

17. This problem combines the operations of squaring a number, determining the square root of a number, and multiplying by a decimal.

The formula for determining the discharge from a nozzle tip is:

$$\text{Dis} = 29.7 \, D^2 \sqrt{P}$$

where   D = nozzle tip diameter

P = nozzle pressure

What is the discharge from the tip shown in Figure 7.7?

SOLUTION HELP—Figure 7.7 shows that the diameter of the tip is 1 inch and the nozzle pressure is 50 psi.

Figure 7.7

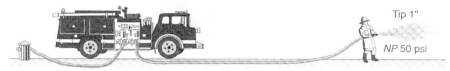

First, square the diameter of the tip and then find the square root of the nozzle pressure.

$$D^2 = (1)(1)$$

The square root of the nozzle pressure is 7.07 (Unit 3 of Chapter 4).
Next, place the results into the formula.

$$Dis = (29.7)(1)(1)(7.07)$$

Now, solve the formula.

$$Dis = (29.7)(1)(1)(7.07)$$
$$= 209.98 \text{ gpm}$$

18. This problem involves the combined operations of squaring a number and multiplying numbers.

    The friction loss in a hose layout is determined by the formula:

    $$FL = CQ^2L$$

    where   C = friction loss coefficient

    Q = flow in hundreds of gpm (flow/100)

    L = hose length in hundreds of feet (length divided by 100)

    The friction loss coefficient for the $\frac{3}{4}$-inch line used in Figure 7.8 is 1100. What is the friction loss in the hose layout?

    SOLUTION HELP—From Figure 7.8, the discharge flow is 25 gpm and the length of the $\frac{3}{4}$-inch booster line is 150 feet.

    First, determine the square of the flow.

    $$.25^2 = (.25)(.25)$$

    Second, place the square of the flow and the length of the hose into the formula.

    $$FL = CQ^2L$$
    $$= (1100)(.25)(.25)(1.5)$$

    Solve the formula.

    $$FL = (1100)(.25)(.25)(1.5)$$
    $$= 103.13 \text{ psi}$$

19. This problem involves the same combined operations as those in Problem 18. What is the friction loss in the hose layout?

    SOLUTION HELP—The friction loss formula is:

    $$FL = CQ^2L$$

Figure 7.8

150' – $\frac{3}{4}$" booster line

Flow 25 gpm

Friction loss = ?

Figure 7.9

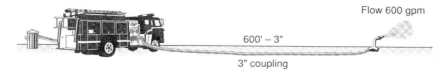

The coefficient for the 3-inch hose shown in Figure 7.9 is 0.677. The flow is 600 gpm. The length of the hose is 600 feet.

First, square the flow divided by 100.

$$(6)^2 = 36$$

Place the coefficient, the square of the flow, and the line length/100 into the formula.

$$FL = (.677)(6)(6)(6)$$

Solve the formula.

$$FL = (.677)(6)(6)(6)$$
$$= 146.23 \text{ psi}$$

20. This problem contains the combination of operations involving the multiplication of decimals and the division of a number.

    There are 2,000 pounds in a ton. To determine the number of tons of water in a cylindrical container, use the formula:

$$\text{tons} = \frac{\text{weight}}{2000}$$

where weight = 62.5 V
       V = volume in cubic feet

Problem: How many tons of water are there in the cylindrical container shown in Figure 7.10?

SOLUTION HELP—From Figure 7.10, the volume is 6,540.7 cubic feet.

First, solve for the weight by using the formula weight = 62.5 V.

$$\text{weight} = 62.5 \text{ V}$$
$$= (62.5)(6,540.7)$$

Place the weight into the formula and solve the formula.

$$\text{tons} = \frac{\text{weight}}{2,000}$$
$$= \frac{(62.5)(6,540.7)}{2,000}$$
$$= \frac{40,8793.75}{2,000}$$
$$= 204.40 \text{ tons}$$

Figure 7.10

Volume 6540.7 cu ft

21. This problem contains the combined operations of squaring a number and multiplying decimals.

    Determining the volume of a cylindrical container can be solved using the formula:

    $$V = .7854 \, D^2 \, H$$

    where V = volume in cubic feet
    D = diameter
    H = height

    Figure 7.11

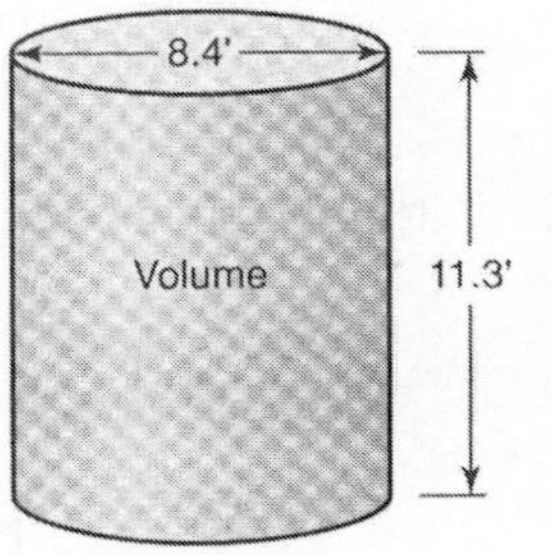

    Problem: What is the volume in Figure 7.11?

    SOLUTION HELP—From Figure 7.11, the diameter is 8.4 feet and the height is 11.3 feet.

    First, square the diameter.

    $$(8.4)^2 = (8.4)(8.4)$$

    Next, place the diameter squared and the height into the formula.

    $$V = (.7854)(8.4)(8.4)(11.3)$$

    Now, solve the formula.

    $$V = (.7854)(8.4)(8.4)(11.3)$$
    $$= 626.22 \text{ cubic feet}$$

22. This problem contains the combined operations of adding whole numbers and decimals and subtracting decimals.

    The required pump discharge pressure can be determined by the formula:

    $$RPDP = NP + FL + BP + AFL + Prepiping - FP$$

    where $RPDP$ = required pump discharge pressure
    NP = nozzle pressure
    FL = friction loss in the hose
    BP = back pressure
    AFL = appliance friction loss
    Prepiping = prepiping friction loss
    FP = forward pressure

    Problem: Using the figures shown in Figure 7.12, determine the required pump discharge pressure.

    SOLUTION HELP—From Figure 7.12, the nozzle pressure is 80, the friction loss is 42.3, the back pressure is 20.4, the forward pressure is 52.7, the appliance friction loss is 25, and the prepiping friction loss is 10.

    First, place all the knowns into the formula.

    $$RPDP = NP + FL + BP + AFL + Prepiping - FP$$
    $$= 80 + 42.3 + 20.4 + 25 + 10 - 52.7$$

    Now, solve the formula.

    $$RPDP = 80 + 42.3 + 20.4 + 25 + 10 - 52.7$$
    $$= 125 \text{ psi}$$

Figure 7.12

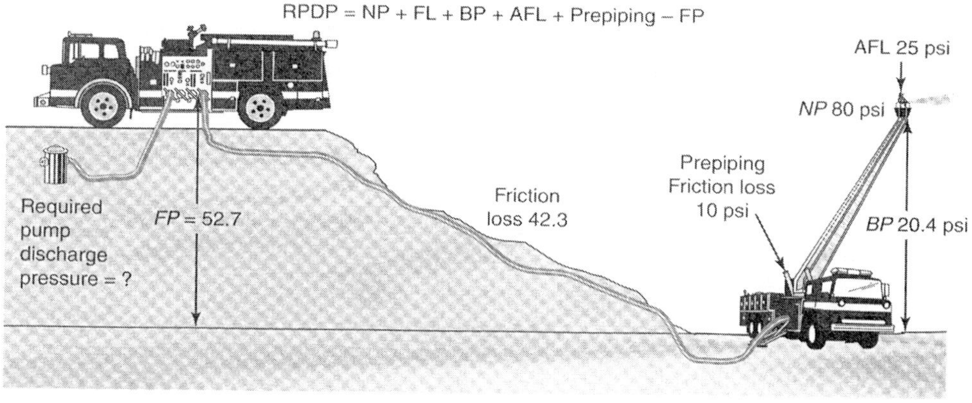

$RPDP = NP + FL + BP + AFL + Prepiping - FP$

- AFL 25 psi
- NP 80 psi
- Prepiping Friction loss 10 psi
- BP 20.4 psi
- Friction loss 42.3
- FP = 52.7
- Required pump discharge pressure = ?

23. The firefighting aircraft shown in Figure 7.13 carries 3,200 gallons of water in its drop tank. By refilling, it completed twelve drops during the day.

    Problem: How many tons of water did it drop on the fire during the day?
    SOLUTION HELP—Multiply 12 × 3200 = 38,400 gallons.
    Then divide 38,400 by 2,000 = 19.2 tons

Figure 7.13

Courtesy of Rick McClure, LAFD

Chapter Summary

This chapter explained how to solve mathematical problems involving more than a single operation, and described the standard procedure for handling the order of operations. The test at the end of the chapter then included an extensive number of problems for the reader to solve.

CHAPTER 8
UNDERSTANDING MEASUREMENTS

KEY TERMS

area	force	pressure
back pressure	forward pressure	rectangle
body heat	friction loss	required pump discharge
capacity	head	pressure
centi-	hecto-	residual pressure
centigrade	inch	square
circle	kilo-	static pressure
deca-	kilometer	systolic pressure
deci-	length	temperature
diastolic pressure	mass	triangle
Fahrenheit	meter	volume
fever	mile	weight
flow pressure	milli-	yard
fluid	pound	
feet	prepiping	

OBJECTIVES AND GOALS

Upon completing this chapter, the reader should be able to:

- List the standard prefixes for the metric system.
- List the standard units in the metric system for length, volume, and mass.
- Provide the basic unit for length in the metric system.
- Provide the formulas for determining the amount of water required for extinguishing and controlling a structure fire as provided by the National Fire Academy.
- Provide the formulas for determining back pressure and forward pressure, and for determining the friction loss in hose lines.
- Provide the formulas for determining the required pump discharge pressure, and for determining the volume of both rectangular and cylindrical containers.

- Determine the weight of water in various storage units, and for determining the number of gallons of water in a hose line.
- Convert Fahrenheit temperature to Centigrade temperature, and Centigrade temperature to Fahrenheit temperature.
- Define normal blood pressure, force, static pressure, flow pressure, residual pressure, pressure as related to head, volume, fluid, and temperature.
- Explain the upper and lower numbers related to blood pressure.
- Work problems related to the formulas provided throughout the chapter.

INTRODUCTION

Only two primary systems of measurement are used throughout the world today: the metric system and the customary system. The customary system is officially used in only the United States and two other countries, Liberia and Myanmar (formerly Burma). It is the primary system used throughout this book; however, the metric system is referred to in regards to EMS operations.

Many individuals will argue that the metric system is much easier and more practical to use than the customary system. In fact, a number of attempts have been made to have it adopted as the official system in the United States; however, these attempts have always failed. It is interesting to note that the metric system missed becoming naturalized in the United States by only one vote in the Continental Congress in the late 1700s or early 1800s. The metric system, however, was adopted by the scientific community in this country almost from its inception. For example, currently the U.S. penny weighs exactly 2.5 grams and the U.S. nickel weighs exactly 5 grams. Both systems are used in the emergency services in the United States.

One of the basic arguments put forward that the metric system is easier to use is that it uses standard prefixes (milli, centi, deci, kilo, hecto, and deca) and a single basic unit for each type of measurement. It uses meter for length, liter for volume, and gram for mass. Designations of multiples and subdivision of any unit may be determined by combining the name of the unit with one of the standard prefixes:

deca means 10

hecto means 100

kilo means 1,000

At the other end:

deci means one-tenth

centi means one-hundredth

milli means one-thousandth

To change a prefix to the number of units of the basic unit, multiply the prefix by the numbers shown in the following chart.

Prefix	Multiply by
deca-	10
hecto-	100
kilo-	1,000
deci-	0.1
centi-	0.01
milli-	0.001

As an example, the basic unit for length measurements is meter. The use of the prefix with this unit are:

decameter	One decameter equals 10 meters
hectometer	One hectometer equals 100 meters
kilometer	One kilometer equals 1,000 meters
decimeter	One decimeter equals $\frac{1}{10}$ of a meter
centimeter	One centimeter equals $\frac{1}{100}$ of a meter
millimeter	One millimeter equals $\frac{1}{1000}$ of a meter

In contrast, the customary system uses different units for each type of measurement: length, volume, and weight. However, there is one type of measurement that is the same for both the metric system and the customary system: the measurement of time. For example, Canada uses the metric system whereas the United States uses the customary system. However, the number of days in a year, the number of hours in a day, the number of minutes in an hour, and the number of seconds in a minute are the same in both countries.

In the customary system, it is generally considered that there are three forms of measurement: *length*, which uses the *inch* as the basic unit; *mass*, which uses the basic unit of an ounce; and *capacity*, which also uses the basic unit of an ounce.

Because the emergency services use a number of different types of measurements, the measurement areas in this book have been expanded to include:

Length measurements

Area measurements

Pressure measurements

Volume measurements

Weight measurements

Fluid measurements

Temperature measurements

Length Measurements

METRIC SYSTEM

The *meter* is the basic unit for length. The word meter is sometimes spelled metre, primarily in Britain. However, both meter and metre are pronounced the same. The following charts show the equivalencies between the units of measurement.

1 kilometer (km)	=	1,000 meters		
1 hectometer (hm)	=	100 meters		
1 dekameter (dam)	=	10 meters		
basic unit	=	1 meter		
10 decimeters (dm)	=	1 meter	=	(1/10 of a meter)
100 centimeters (cm)	=	1 meter	=	(1/100 of a meter)
1,000 millimeters (mm)	=	1 meter	=	(1/1,000 of a meter)
10 millimeters (mm)	=	1 centimeter (cm)		
10 centimeters	=	1 decimeter (dm)	=	100 millimeters
10 decameters	=	1 meter (m)	=	1,000 millimeters

10 meters	=	1 decimeter (dam)		
10 decameters	=	1 hectometer (hm)	=	100 meters
10 hectometers	=	1 kilometer (km)	=	1,000 meters

CUSTOMARY SYSTEM

The inch is the basic unit of length for the customary system. The following are additional units for measurement together with their equivalents.

feet (ft)	=	12 inches
yard (yd)	=	3 feet
mile (mi)	=	5,280 feet or 1,760 yards

MEASUREMENT CONVERSIONS

The following chart can be used to convert length units between the customary and metric systems.

Customary System	Metric System
1 inch	2.54 cm
1 feet	30.48 cm
1 yard	91.44 cm
1 mile	1,609 meters
3,281 feet	1 km

The following chart is also useful for making conversions.

To Convert	to	Multiply by
meters	inches	39.37
inches	meters	.0254
meters	feet	3.28
feet	meters	.305
meters	yards	1.09
yards	meters	.914
kilometers	miles	.62
miles	kilometers	1.609

A Word of Caution Regarding Metric Usage

Metric unit names are ordinary words. However, care should be used with unit symbols. It should be noted that unit symbols

- are sensitive regarding upper and lower cases. Upper and lower case letters have different meanings. As an example, mm is the symbol for millimeter, which is one-thousandth of a meter. However, Mm is the symbol for megameter, which is one million meters.

- do not have singular and plural forms. There is no "s" at the end of any symbol.
- are not abbreviations. Unless it falls at the end of a sentence, there is no period after a symbol.

It is relatively easy to convert units using the above chart for measurement conversions.

Problem: How many centimeters are there in three yards?

Solution: Because there are 91.44 cm in a yard, there are 274.32 cm in three yards ($3 \times 91.44 = 274.32$).

Problem: How many feet are there in three km? How many miles would this equal?

Solution: Because there are 3,281 feet in a km, there are 9,843 feet in 3 km ($3 \times 3,281 = 9,843$). There are 5,280 feet in a mile; the 9,843 feet can be divided by the 5,280 to determine how many miles would equal 3 km ($9,843 \div 5,280 = 1.86$ miles).

The procedure is even easier when working solely in the metric system. Use the following information as a guide.

decameter	=	10 to +1
hectometer	=	10 to +2
kilometer	=	10 to +3
decimeter	=	10 to −1
centimeter	=	10 to −2
millimeter	=	10 to −3

The numbers 1, 2, and 3 refer to how many places the decimal point is moved to convert a unit to meters. The plus numbers indicate that the decimal point is moved to the right. The negative numbers indicate that the decimal point is moved to the left.

Problem: Convert 6,250 millimeters to meters.

Solution: Millimeters has a value of −3. This means that the decimal point should be moved 3 places to the left to convert millimeters to meters.
The solution is:

$$6{,}250 \text{ millimeters} = 6.25 \text{ meters}$$

Problem: Convert 5.15 hectometers to meters.

Solution: Hectometers has a value of +2. This means that to convert hectometers to meters, move the decimal point 2 places to the right.
The solution is:

$$5.15 \text{ hectometers} = 515 \text{ meters}$$

CUSTOMARY LENGTH PROBLEMS

Many customary length problems are given using whole number lengths only. Others are given using mixed numbers. When working with addition or subtraction problems involving mixed numbers, it is important to get the problem set up correctly. Use the following problem as a guide.

Problem: Add 7 feet 5 inches and 3 feet 4 inches.

Solution:

Step 1: Set up the problem, separating the feet in one unit and the inches in a separate unit. Line up the numbers in each of the units.

Feet	Inches
7	5
3	4

Step 2: Add each of the units.

Feet	Inches
7	5
3	4
10	9

Step 3: If the amount of the inches unit is less than 12, combine the feet with the inches. This is the completed answer.

10 feet 9 inches

Try another one.

Problem: Add 9 feet 8 inches + 7 feet 6 inches + 10 feet 5 inches

Step 1: Set up the problem, separating the feet in one unit and the inches in a separate unit. Line up the numbers in each of the units.

Feet	Inches
9	8
7	6
10	5

Step 2: Add each of the units.

Feet	Inches
9	8
7	6
10	5
26	19

Step 3: If the number of inches is 12 or more, remove 12 of them and add a 1 to the feet unit. Keep the remainder in the inches unit.

Feet	Inches
9	8
7	6
10	5
26	19
+1	−12
27	7

Step 4: Combine the feet with the inches. This is the completed answer to the problem.

27 feet 7 inches

Subtraction problems are handled in a similar fashion.

Problem: Subtract 3 feet 3 inches from 8 feet 7 inches.

Solution:

Step 1: Set up the problem, separating the feet in one unit and the inches in a separate unit. Line up the numbers in each of the units.

Feet	Inches
8	7
3	3

Step 2: Subtract each of the units.

Feet	Inches
8	7
−3	−3
5	4

Step 3: If the number of inches is less than 12, combine the feet with the inches. This is the completed answer.

5 feet 4 inches

Try another one.

Problem: Subtract 5 feet 9 inches from 8 feet 2 inches.

Step 1: Set up the problem, separating the feet in one unit and the inches in a separate unit. Line up the numbers in each of the units.

Feet	Inches
8	2
−5	−9

Step 2: If the number of inches in the subtrahend is greater than the number of inches in the minuend, take a feet from the feet unit (it is 12 inches) and add the 12 inches to the inches in the minuend.

Feet	Inches
8 − 1 = 7	2 + 12 = 14
5	9

Step 3: Subtract each of the units.

Feet	Inches
8 − 1 = 7	2 + 12 = 14
−5	−9
2	5

Step 4: Combine the feet with the inches. This is the completed answer to the problem.

2 feet 5 inches

Reader's Practice Problems—8.1

Feel free to use the conversion charts previously introduced when solving the following conversion problems.

1. Convert 5 meters to inches.
2. Convert 429 feet to meters.
3. Convert 6 kilometers to miles.
4. Convert 21 miles to kilometers.
5. Convert 3 meters to kilometers.
6. Convert 32 centimeters to meters.
7. Add 8 feet 6 inches and 3 feet 3 inches.
8. Add 9 feet 9 inches and 13 feet 9 inches.
9. Subtract 4 feet 2 inches from 9 feet 7 inches.
10. Subtract 5 feet 10 inches from 12 feet 2 inches.

The answers are found in the Reader's Practice Problems section of the Appendix.

LENGTH REVIEW QUESTIONS

1. What are the six standard prefixes used in the metric system?
2. What is the basic unit for length in the metric system?
3. How many meters are there in a kilometer?
4. How many millimeters are there in a meter?
5. What is the basic unit for length in the customary system?
6. How many cm are there in an inch?
7. How many meters are there in a mile?
8. What do you multiply by to change meters to inches?
9. What do you multiply by to change feet to meters?
10. What do you multiply by to change yards to meters?
11. What are the three principles that should be noted regarding metric unit symbols?
12. How should the problem be set up when working with addition problems involving mixed numbers in the customary system?

Test 8.1

1. How many feet are there in $1\frac{1}{2}$ meters?
2. Convert 3,752 millimeters to meters.

Figure 8.1

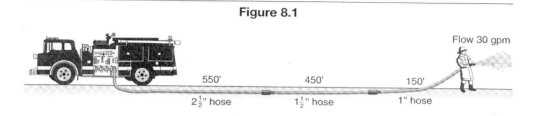

3. Convert 17 kilometers to meters.
4. Convert 82 decimeters to meters.
5. How many feet are there in 3 miles?
6. How many inches are there in 50 feet?
7. Add 9 feet 3 inches and 5 feet 6 inches.
8. Add 11 feet 8 inches and 7 feet 9 inches.
9. Subtract 2 feet 8 inches from 9 feet 11 inches.
10. Subtract 5 feet 10 inches from 9 feet 3 inches.
11. How many yards are there in 17 meters?
12. How many kilometers are there in 3.2 miles? Round your answer to the second decimal point.
13. How many meters are there in 127 inches? Round your answer to the second decimal point.
14. How many feet are there in 9.7 meters? Round your answer to the second decimal point.
15. How many inches are there in 7.2 meters? Round your answer to the second decimal point.
16. How many meters are there in 189.2 feet? Round your answer to the second decimal point.
17. How many miles are there in 79 kilometers? Round your answer to the second decimal point.
18. What is the total length of the line from the pumper to the firefighter holding the nozzle in Figure 8.1?

    SOLUTION HELP—Add the length of the $2\frac{1}{2}$-inch hose, the $1\frac{1}{2}$-inch hose, and the 1-inch hose.

19. How much longer is the $2\frac{1}{2}$-inch hose in Figure 8.2 than the 1-inch hose?

    SOLUTION HELP—Subtract the 1-inch hose from the $2\frac{1}{2}$-inch hose.

Figure 8.2

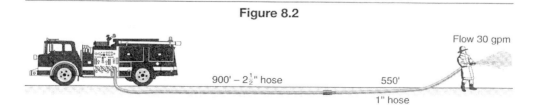

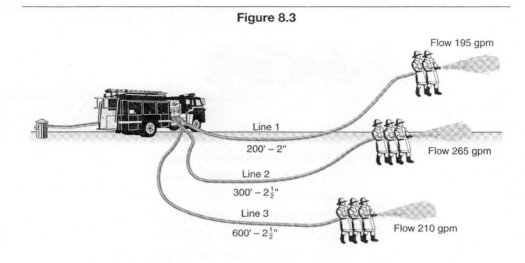

Figure 8.3

Figure 8.4

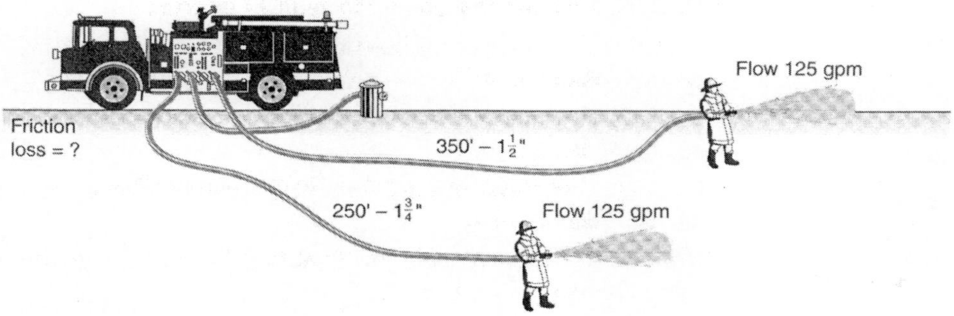

20. In the layouts shown in Figure 8.3, how much longer are the combined lengths of lines 2 and 3 than the length of line 1?

    SOLUTION HELP—Add the lengths of line 2 and line 3, then subtract the length of line 1 from the result.

21. How many more inches in length is the $1\frac{1}{2}$-inch line shown in Figure 8.4 than the $1\frac{3}{4}$-inch line?

    SOLUTION HELP—Subtract the length of the $1\frac{3}{4}$-inch hose from the $1\frac{1}{2}$-inch, then multiply the answer by 12.

22. Including the 4-inch supply line, how much hose has been laid out in the configuration shown in Figure 8.5?

    SOLUTION HELP—Add the length of the supply line, the length of the 2-inch hose, and the length of the $1\frac{1}{2}$-inch hose.

23. Another 450 feet of $1\frac{3}{4}$-inch hose has been ordered to be taken off the pumper shown in Figure 8.6. When this layout is completed, how much hose will be laid off the pumper? (do not count the supply line)

    SOLUTION HELP—Add the length of the $1\frac{1}{2}$-inch hose, the length of the $1\frac{3}{4}$-inch hose, and the 450 feet addition.

Figure 8.5

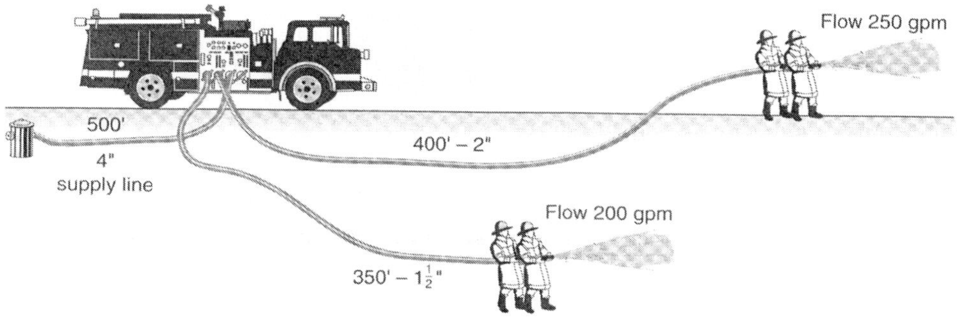

Figure 8.6

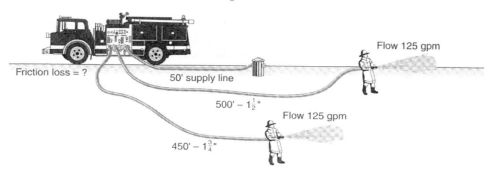

24. It should be noted that the nozzle on line 1 has been shut down. Prior to the firefighters shutting off the nozzle on line 1, the pumper in Figure 8.7 was supplying water to a total of 1,200 feet of hose to the three attack lines. How long is line 1?

SOLUTION HELP—Add the length of line 3 and the length of line 2, and subtract the result from 1,200.

Figure 8.7

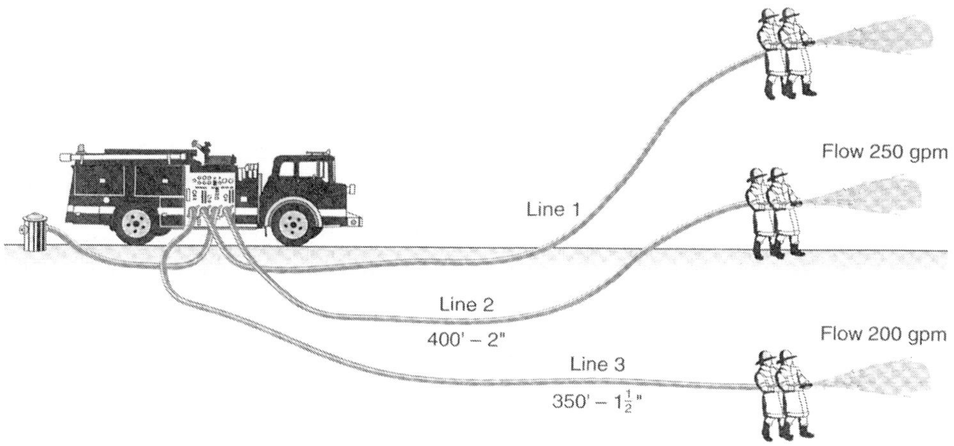

Figure 8.8

25. What is the equivalent length of the attack line in Figure 8.8 in centimeters?
SOLUTION HELP—Multiply the length of the 1½-line × 30.48.

Area Measurements

An *area* measurement is a two-dimensional measurement that measures the surface of an area. In the fire service, the areas most often considered are squares, rectangles, triangles, and circles. Each of these use a formula to determine the area. The area of these surfaces are generally expressed in square inches or square feet. These measurements can be converted to square centimeters or square meters.

The approximate conversions from customary area measurements to metric area measurements are:

To convert square inches (in^2) to square centimeters (cm^2) multiply by 6.5.

To convert square feet (ft^2) to square meters (m^2) multiply by 0.09.

The approximate conversions from metric area measurements to customary area measurements are:

To convert square centimeters (cm^2) to square inches (in^2) multiply by 0.16.

To convert square meters (m^2) to square feet (ft^2) multiply by 10.76.

SQUARES

A *square* is a two-dimensional geometric figure having four right angles and four equal sides (see Figure 8.9).

The formula for finding the area of a square is:

$$A = S^2$$

where A = area
S = length of side

Figure 8.9

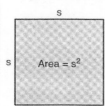

Problem: What is the area of a square that has 3-feet sides?

Solution:

$$A = 3' \times 3'$$
$$= 9 \text{ square feet}$$

Problem: What would this equal in square meters?

Solution: To convert 9 square feet to square meters, multiply by 0.09.

$$9 \times 0.09 = 0.81$$

Problems solving for a square only are not generally included in the hydraulics area of the fire service. However, a container that has a one-square-inch base and is one feet high will hold water that weighs .434 pounds. This information will be included in a number of formulas that will be introduced in the next section. Watch for them.

Figure 8.10

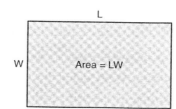

RECTANGLES

A *rectangle* is an oblong plane surface with parallel sides and four right (90°) angles (see Figure 8.10). The formula for determining the area of a rectangle is:

$$A = l \times w$$

where A = area
  l = length
  w = width

Problem: What is the area of a rectangle whose length is 5 meters and width is 3 meters?
Solution:

$$A = 5\,m \times 3\,m$$
$$= 15 \text{ square meters}$$

The National Fire Academy uses a rectangle as a portion of their formula for determining the amount of water required for extinguishing and controlling a structure fire. The formula is:

$$\text{Required fire flow} = \frac{l \times w}{3}$$

Following is a sample of the use of the formula.

Problem: The first-in officer to a reported building fire estimated that the area of involvement of the building measured 40 feet by 30 feet. How much water is estimated as being needed to control and extinguish this fire? (answer is in gallons per minute [gpm])
Solution:

$$\text{Required fire flow} = \frac{l \times w}{3}$$

where l = 40 feet
  w = 30 feet

$$\text{Required fire flow} = \frac{40 \times 30}{3} = \frac{1{,}200}{3} = 400 \text{ gpm}$$

TRIANGLES

Figure 8.11

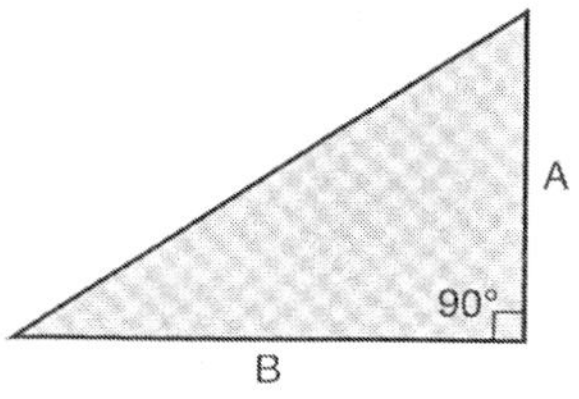

A *triangle* is a two-dimensional geometric figure having three sides and three angles (see Figure 8.11). If one of the angles is a right angle, it is called a right-triangle. The formula for determining the area of a right-triangle is:

$$\text{Area} = \frac{ab}{2}$$

where a = altitude (height)
  b = base

Problem: What is the area of a right-triangle that has an altitude of 8 inches and a base of 12 inches?

Solution:
$$\text{area} = \frac{(8")(12")}{2} = \frac{96}{2} = 48 \text{ square inches}$$

Problem: What would this area equal in square centimeters?

Solution: To convert square inches to square centimeters, multiply by 6.5.

$$48 \times 6.5 = 312 \text{ square centimeters}$$

CIRCLES

A *circle* is a two-dimensional geometric figure that is formed by a curved line surrounding a center point with every point on the formed line being an equal distance from the center point (see Figure 8.12). The formula for determining the area of a circle is:

$$A = \pi r^2$$

where A = area
 π = 3.1416
 r = radius

(*Note:* The radius is $\frac{1}{2}$ the diameter.)

or

$$A = .7854 \, D^2$$

where d = diameter

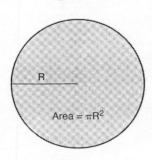

Figure 8.12

Problem: What is the area of a circle that has a diameter of 7 feet?

Solution:
$$A = (3.1416)(3.5)(3.5) = 38.4846 \text{ square feet}$$

or

$$A = (.7854)(7)(7) = 38.4846$$

Note: The formula commonly used in the fire service is: Area = .7854 D^2. This probably came about because the size of the nozzle tip on smooth-bore tips is given in the diameter of the tip in inches.

Reader's Practice Problems—8.2

Feel free to use the conversion information introduced earlier in this section.

1. What is the area of a square that has 9-inch sides?
2. What is the equivalent number of centimeters for this square?
3. A firefighter has a home that is situated on a lot that measures 200 feet by 200 feet. How many square feet are on the lot?
4. What is the size of the lot in meters?
5. How many square meters are there in a rectangle that has a length of 6 meters and a width of 4.5 meters?
6. The first-in officer estimates that a 20-by-25-feet section of the 1st floor of a large warehouse is involved with fire. How much water flow would be required to control and extinguish this fire?
7. How many square feet are there in a triangle that has a base of $4\frac{1}{2}$ feet and a height of 3 feet?
8. How many square inches are there in a 2-inch tip on the end of a hose line?

9. What is this equivalent to in centimeters?
10. How many square inches are there on a half-inch tip of a nozzle located at the end of a 1-inch line?

The answers are found in the Reader's Practice Problems section of the Appendix.

AREA REVIEW QUESTIONS

1. What is the definition of area?
2. What is the definition of a square?
3. What formula is used for determining the area of a square?
4. What is the definition of a rectangle?
5. What formula is used to determine the area of a rectangle?
6. What is the National Fire Academy's formula for determining the required fire flow for the interior of a structure?
7. What is the definition of a triangle?
8. What formula is used to determine the area of a right triangle?
9. What is the definition of a circle?
10. What formula is commonly used in the fire service for determining the area of a circle?
11. What should a number of square inches be multiplied by to convert them to an approximate number of square centimeters?
12. What should a number of square feet be multiplied by to convert them to an approximate number of square meters?
13. What should a number of square centimeters be multiplied by to convert them an approximate number of square inches?
14. What should a number of square meters be multiplied by to convert them to an approximate number of square feet?

Test 8.2

1. What is the area of a square that has a side of 3 millimeters?
2. What is the area of a rectangular that has a length of 35 inches and a width of 16 inches?
3. What is the area of a right triangle that has a height of 9 feet and a base of 12 feet?
4. What is the area of a circle that has a diameter of 3.5 meters?
5. The Incident Commander at a structure fire estimates that the area of involvement measures 60 feet by 400 feet. How much water would be required for extinguishing and controlling this fire?
6. The area of an acre of land is 4,840 square yards. What would be the length of each side of a square acre?
7. What is the area of a rectangle that has a length of $8\frac{1}{3}$ feet and a height of $5\frac{1}{4}$ feet?
8. A two-story single-family residence measures 52 × 41 feet on the first floor and 37 × 33 feet on the second floor. How many square feet is the house?
9. The distance across a round bull ring is 72 feet. What is the area inside the bull ring?
10. Upon arrival, the Incident Commander determined that approximately $\frac{2}{3}$ of a 3,800-square feet house was involved with fire. What would be the needed fire flow for this fire if there are no external exposures?

11. A firefighter owns a piece of ground that measures 200 feet × 200 feet. What is the area of this land?
12. What would be the area of a rectangle that measures 5 feet by 6 feet?
13. According to the National Fire Academy, what amount of water should be used on the interior of a fully involved 125-by-200-feet single-story isolated commercial building?
14. What is the length of the hypotenuse of a triangle that has a base of 6 feet and a height of 5 feet?
15. What is the area in square inches of a solid-bore 3-inch tip used on a portable monitor?

Pressure Measurements

Pressure is generally thought of as force applied over a given area. In the fire service, the area normally used is a square inch. The result is expressed as pounds per square inch (psi) and is used in the deliverance of water. However, pounds per square feet is used in computing figures for some small storage containers. In the *metric* system, the pressure is normally expressed in Newtons per square meter (pascals).

Two areas of pressure are important to the fire service and the EMS. The fire service is interested in pressure as it relates to water; the EMS is interested in pressure as it relates to blood.

BLOOD PRESSURE

Almost every individual has had his or her blood pressure taken at one time or another. The results are normally given in two numbers and sometimes three. The two numbers are generally referred to as if they were one; for example, 120 over 80. He or she might also be given a pulse rate, which is the number of times the heart beats in one minute.

The higher of the two numbers refers to the blood pressure when the heart is supplying blood to the body. It is called the ***systolic pressure***. The lower number is the pressure of the blood when the heart is at rest. It is called the ***diastolic pressure***.

Strange as it may seem, regardless of how many times an individual has had his or her blood pressure taken, other than medical personnel, very few individuals know what unit is used for measuring blood pressure. The unit is millimeters of mercury.

Blood pressure results are divided into a number of different categories. The categories for adults measured in mmHg (millimeters of mercury) are shown in the following chart.

Category	Systolic (top number)	Diastolic (bottom number)
Normal	Less than 120	Less than 80
Prehypertension	120–139	80–89
High blood pressure		
Stage 1	140–159	90–99
Stage 2	160 or higher	100 or higher

A couple of points should be remembered in regards to the above chart.

1. If the systolic and diastolic blood pressures are in different categories, the one in the higher category should be used for classifying an individual's blood pressure.
2. If an individual has diabetes or chronic kidney disease, a blood pressure of $\frac{130}{80}$ or higher is considered as high blood pressure.

PRINCIPLES OF PRESSURE IN WATER

Although the EMS works with pressure in the metric system, the fire service works with pressure in the customary system. The primary types of pressure measurements involve pressure as it is related to water. The first thing that should be remembered when considering the measurement of water pressure is the difference between force and pressure.

Force refers to the amount of energy that is being applied at a given point, or points. It is usually expressed in pounds.

Pressure is usually defined as force per unit area. It may be considered as the measurement of energy in water and is usually expressed in pounds per square inch (psi). Several types of pressure are related to water.

- *Static pressure* is the pressure of water when it is not in motion. An example would be the pressure obtained at a pressure gauge on an outlet of a hydrant when no water is flowing from the hydrant.
- *Flow pressure* is the pressure taken after the water has been placed in motion. The pressure at the outlet of a nozzle when water is flowing is a good example of flow pressure. This pressure is referred to as nozzle pressure.
- *Residual pressure* is the pressure remaining at a hydrant outlet once the water is flowing.

Several formulas are used in the fire service that measure pressure. The most common is:

$$P = .434 H$$

where P = pressure
        H = head

Head is defined as the vertical distance from the surface of the water to the point being considered. An example is shown in the following problem.

Problem: What is the pressure at the hydrant in the situation shown in Figure 8.13?

Solution:

$$P = .434 H$$
$$H = 145 \text{ feet}$$

Figure 8.13

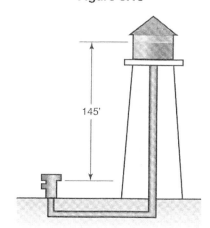

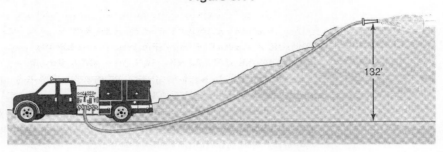

Figure 8.14

then
$$P = (.434)(145)$$
$$= 62.93 \text{ psi (static pressure)}$$

Another pressure formula is:

$$BP = .434 \, H$$

where   BP = back pressure
        H = 120 feet

then

$$BP = (.434)(120)$$
$$= 52.08 \text{ psi}$$

Back pressure is the pressure working back against the pump in the apparatus when the water is being supplied at a level above the pumper outlet. An example is illustrated in the following problem.

Problem: What pressure does the pumper have to provide to overcome the back pressure when water is being discharged from the nozzle tip shown in Figure 8.14?

Solution:

$$BP = .434 \, H$$
$$H = 132 \text{ feet}$$

then

$$BP = (.434)(132)$$
$$= 57.29 \text{ psi}$$

Forward pressure is a situation where pressure aid is given to a water source such as a pump when the water being used is below the water source. The formula for determining the forward pressure is the same as that for back pressure except that an FP is used instead of a BP. An example is illustrated in the following problem.

Problem: The pumper shown in Figure 8.15 is pumping to a nozzle that is located below the level of the pumper. What is the forward pressure in this situation?

Solution:

$$FP = .434 \, H$$

where   H = 116 feet

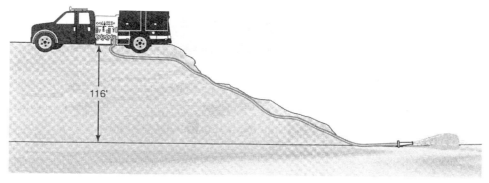

Figure 8.15

then

$$FP = (.434)(116)$$
$$= 50.34 \text{ psi}$$

Friction loss is the amount of pressure lost in a hose line due to water moving through the line. Friction loss is present at every situation where a pumper is supplying water to firefighters operating a nozzle. The formula for determining the amount of the friction loss is:

$$FL = CQ^2L$$

where
- FL = friction loss in the hose line
- C = friction loss coefficient
- Q = flow in hundreds of gpm (flow/100)
- L = hose length in hundreds of feet (length/100)

Note: Following are the friction loss coefficients for single hose lines.

Hose Diameter (in inches)	Coefficient
$\frac{3}{4}$ (booster)	1,100
1 (booster)	150
$1\frac{1}{2}$ (booster)	80
$1\frac{1}{2}$	24
$1\frac{3}{4}$ (with $1\frac{1}{2}$-inch couplings)	15.5
2	8
$2\frac{1}{2}$	2
3 (with $2\frac{1}{2}$-inch couplings)	0.8
3 (with 3-inch couplings)	0.677
$3\frac{1}{2}$	0.34
4	0.2
$4\frac{1}{2}$	0.1
5	0.8
6	0.05

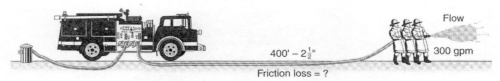

Figure 8.16

An example is illustrated in the following problem.

Problem: What is the friction loss in the 2½-inch line shown in Figure 8.16?

Solution:

$$FL = CQ^2L$$

where C = 2 (taken from the chart just introduced)
Q = 300/100 = 3
L = 400/100 = 4

then FL = (2)(3)(3)(4) = 72 psi

Required pump discharge pressure is the total pressure that a pumper must provide in order to take care of all the pressure losses in a layout plus the pressure required at the nozzle. Following is the maximum requirements for a pumper; however, it is seldom necessary for a pumper to supply all of them. Only those associated with a particular layout are included when determining the required discharge pressure.

$$RPDP = NP + FL + BP + AFL + \text{prepiping} - FP$$

where RPDP = required pump discharge pressure
NP = nozzle pressure
FL = friction loss in the hose
BP = back pressure
AFL = appliance friction loss
Prepiping = piping built into an appliance such as a ladder pipe
FP = forward pressure

In the following situation, the pumper has to provide for the friction loss in the hose, the back pressure, the appliance friction loss, and the nozzle pressure. Only these items are included in the formula.

Problem: What is the RPDP for the situation shown in Figure 8.17?

Solution:

$$RPDP = NP + FL + BP + AFL$$

where NP = 100 psi
FL = 36 psi
BP = 22 psi
AFL = 15 psi

then RPDP = 100 + 36 + 22 + 15 = 173 psi

Figure 8.17

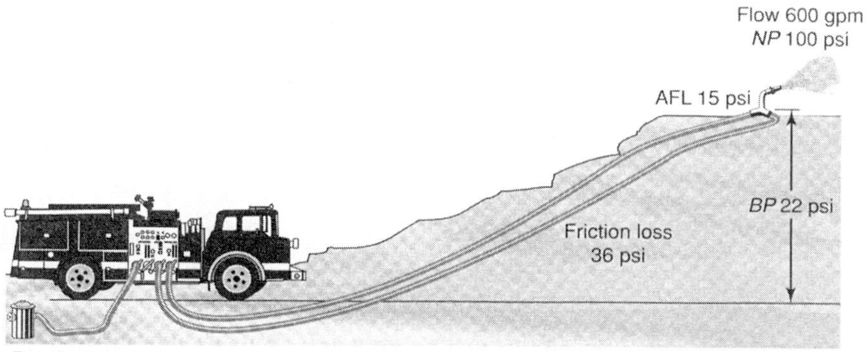

Flow 600 gpm
NP 100 psi
AFL 15 psi
BP 22 psi
Friction loss 36 psi
Required pump discharge pressure = ?

Reader's Practice Problems—8.3

1. A firefighter paramedic takes the blood pressure of a victim of a heart attack. The readings are 141 over 90. What is the difference between the victim's systolic and diastolic pressure?
2. The level of the water serving a hydrant is located 124 feet above the hydrant outlet. What is the static pressure on the hydrant outlet?
3. When the hydrant outlet is opened fully and water is flowing, the loss in pressure is 32 psi. What is the flow pressure?
4. A line has been taken off a pumper and advanced up a hill to be used on a house fire that is located 145 feet above the pumper outlet. What is the back pressure in this situation?
5. A line has been taken off a pumper and the nozzle is being worked on a large trash fire that is located 240 feet below the pumper outlet. What is the forward pressure aiding the pump on this layout?
6. What is the friction loss in a 350-feet length of $2\frac{1}{2}$-inch hose when the flow is 300 gpm? The C factor for a $2\frac{1}{2}$-inch line is 2.
7. The friction loss coefficient (C) for a $1\frac{1}{2}$-inch hose is 24. What is the friction loss in 150 feet of $1\frac{1}{2}$-inch hose if the nozzle is flowing 80 gpm?
8. The nozzle pressure at the tip on a layout of hose is 50 psi. The friction loss is 44 psi and the back pressure is 30 psi. What is the required pump discharge pressure for this situation?

The answers are found in the Reader's Practice Problems section of the Appendix.

PRESSURE REVIEW QUESTIONS

1. What are the two areas of importance to the fire service and the EMS regarding pressure?
2. If an individual's blood pressure is taken and given in two numbers, what is the first number called?

3. What is the second number called?
4. To what does systolic pressure refer?
5. To what does diastolic pressure refer?
6. What is considered a normal blood pressure?
7. What unit is used for measuring blood pressure?
8. What is the systolic number for Stage 2 high blood pressure?
9. Define force.
10. Define static pressure.
11. Define flow pressure.
12. Define residual pressure.
13. What is the common formula used in the fire service to determine pressure produced as a result of gravity?
14. Define head in reference to pressure.
15. What formula is used for determining back pressure?
16. What formula is used for determining the friction loss in hose?
17. What formula is used for determining the required pump discharge pressure?
18. Define back pressure.
19. Define forward pressure.
20. What formula is used for determining forward pressure?

Test 8.3

1. What is the static pressure on the hydrant shown in Figure 8.18?

   SOLUTION HELP—P = .434 H

   H = 467 feet

   then
   P = (.434)(467)
   = 202.68

2. What is the back pressure on the pumper in the situation shown in Figure 8.19?

   SOLUTION HELP—BP = .434 H

   H = 136 feet

   then
   BP = (.434)(136)
   = 59.02

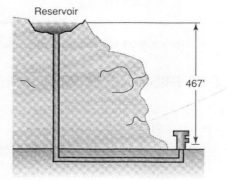

Figure 8.18

Figure 8.19

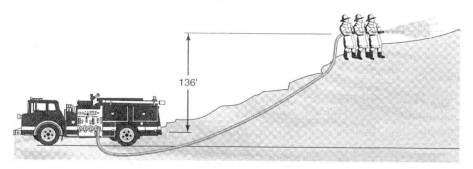

Figure 8.20

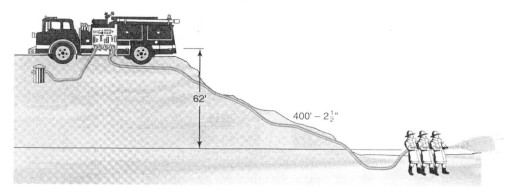

3. What is the forward pressure assisting the pumper in the situation shown in Figure 8.20?

   SOLUTION HELP— FP = .434 H

   where   H = 62 feet

   then $\qquad$ FP = (.434)(62)
   $\qquad\qquad\quad$ = 26.91 psi

4. What is the friction loss in the layout shown in Figure 8.21? The friction loss coefficient (C) for a 2-inch hose is 8.

   SOLUTION HELP— FL = CQ²L

   where   C = 8

   $\qquad$ Q = flow in 100 of gpm (flow/100)

   $\qquad$ L = length in hundreds of feet (length/100)

   then $\qquad$ FL = (8)(1.75)(1.75)(5.50)
   $\qquad\qquad\quad$ = 134.75 psi

Figure 8.21

Figure 8.22

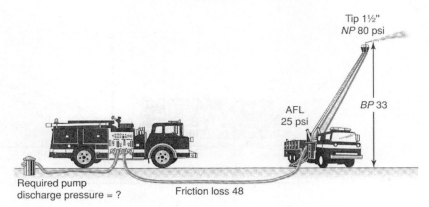

5. What is the RPDP for the layout shown in Figure 8.22?

SOLUTION HELP—RPDP = NP + FL + BP + AFL + prepiping − FP

where
$$NP = 80$$
$$FL = 48$$
$$BP = 33$$
$$AFL = 25$$
$$prepiping = 0$$
$$FP = 0$$

then
$$RPDP = 80 + 48 + 33 + 25$$
$$= 186 \text{ psi}$$

The following situation is used to solve Problems 6 through 10.

Paramedic Unit #7 responded to a vehicle accident in which three individuals were injured. After taking the vitals on each individual, they determined that patient A had a blood pressure of 124 over 82, patient B had 119 over 79, and patient C had 132 over 91.

6. Which of the patients had a blood pressure within normal limits?

7. What was the spread between the systolic and diastolic pressures of patient A?

8. What was the total amount of the systolic pressures for the three patients?

9. What was the total amount of the diastolic pressures for the three patients?

10. What was the average systolic pressure of the three patients?

11. A hydrant outlet is located 112 feet below the surface of the water in the reservoir supplying it. What is the pressure at the outlet when no water is flowing from the hydrant? (answer to second decimal point)

SOLUTION HELP— $P = .434 H$

where $H = 112$ feet

then
$$P = (.434)(112)$$
$$= 48.61 \text{ psi}$$

12. The pressure at the outlet when water is flow is 17 psi in problem 11. What is the difference between static pressure and the flow pressure of the outlet?

SOLUTION HELP—Subtract the flow pressure (17) from the static pressure (48.61).

13. A pumper is supplying water to a nozzle that is located 45 feet above the pump. What is the back pressure working against the pump?

    SOLUTION HELP—BP = .434 H

    where   H = 45

    then
    $$BP = (.434)(45)$$
    $$= 19.53 \text{ psi}$$

14. A pumper is supplying water to a nozzle that is located 132 feet below the pump. What is the forward pressure assisting the pump? (answer to the second decimal point)

    SOLUTION HELP—FP = .434 H

    where   H = 132

    then
    $$FP = (.434)(132)$$
    $$= 57.29 \text{ psi}$$

15. A pumper is supplying 300 gpm through 400 feet of $2\frac{1}{2}$-inch hose. What is the friction loss in this layout? The C for a $2\frac{1}{2}$-inch hose is 2.

    SOLUTION HELP—FL = CQ²L

    where   C = 2 (taken from the chart just introduced)
    $$Q = \frac{300}{100} = 3$$
    $$L = \frac{400}{100} = 4$$

    then $$FL = (2)(3)(3)(4) = 72 \text{ psi}$$

16. The upper level of the water tank that is supplying water to a hydrant is 133 feet. What is the static pressure on the hydrant?

    SOLUTION HELP— Use the formula: P = .434 H. Multiply .434 × 133.

17. What is the back pressure against a pump that is supplying water to a rubbish fire that is located 164 feet above the pump?

    SOLUTION HELP— Use the formula: BP = .434 H. Multiply .434 × 164.

18. What is the friction loss in 550 feet of $2\frac{1}{2}$-inch hose when the flow is 195 gpm?

    SOLUTION HELP—Use the formula: FL = CQ²L. The hose coefficient is 2. Multiply 2 × 1.95 × 1.95 × 5.5.

19. The layout is 450 feet of 2-inch hose with a tip having a nozzle pressure of 50 psi. The friction loss in the line is 94 psi. The line is laid at ground level. What should the pump discharge pressure be?

    SOLUTION HELP—Use the formula: RPDP = NP + FL. Add 50 + 94.

20. The $1\frac{1}{2}$-inch taken off Engine 6 is working on a small fire in a dwelling that is located 144 feet below the level of the pump. What pressure advantage does the pump have on this layout?

    SOLUTION HELP—Use the formula: FP = .434 H. Multiply .434 × 144.

Volume Measurements

INTRODUCTION

Volume refers to the size of a three-dimensional space that is enclosed within an object or occupied by an object. The size of a volumetric container associated with the fire service is normally expressed in cubic inches or cubic feet. The standard unit of volume in the metric system is the liter. One liter is equal to 1,000 cubic centimeters. A liter is a little more than one quart. The other units and their equivalents are:

   1 milliliter (ml) = 0.001 liter
   1 centiliter (cl) = 0.01 liter
   1 deciliter (dl) = 0.1 liter
   1 kiloliter (kl) = 1000 liters

The approximate conversions of customary volume measurements to metric volume measurements are:

To convert cubic inches to cubic centimeters, multiply by 16.39.
To convert cubic feet to cubic meters, multiply by 0.03.
To convert fluid ounces to milliliters, multiply by 30.
To convert gallons to liters, multiply by 3.8.

The approximate conversions of metric volume measurements to customary volume measurements are:

To convert cubic centimeters to cubic inches, multiply by 0.06.
To convert cubic meters to cubic feet, multiply by 35.31.
To convert milliliters to fluid ounces multiply by 0.03.
To convert liters to gallons, multiply by 0.26.

The two most common types of containers found in the fire service that are normally used for the storage of water are rectangular containers and cylindrical containers.

Rectangular Containers

Rectangular containers are normally used as water tanks on apparatus. The formula for determining the volume of a rectangular container is:

$$V = l\,w\,h$$

where V = volume
  l = length
  w = width
  h = height

Following is an example of the application of this formula.

Problem: What is the volume of the apparatus water tank shown in Figure 8.23?

Solution:

$$V = l\,w\,h$$

where $l = 6$
  $w = 3$
  $h = 3$

$$V = (6)(3)(3) = 54 \text{ cubic feet}$$

Figure 8.23

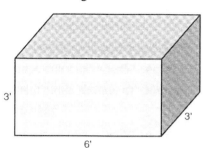

Problem: What would this equal in cubic meters?

Solution: To convert cubic feet to cubic meters, multiply by 0.03.

$$54 \times 0.03 = 1.62 \text{ cubic meters.}$$

Occasionally, it is desirable to convert cubic feet to cubic inches. A cubic feet measures 12 inches $\times$ 12 inches $\times$ 12 inches and therefore contains 1,728 cubic inches.

Problem: The water tank on a fire apparatus has a length of 5 feet, a width of 3.5 feet, and a height of 2.5 feet. What is the volume of this water tank?

Solution:

$$V = l\,w\,h$$

where l = 5
       w = 3.5
       h = 2.5

then
$$V = (5)(3.5)(2.5)$$
$$= 43.75 \text{ cubic feet}$$

Problem: What would this equal in cubic inches?

Solution: One cubic feet equals 1,728 cubic inches.
Therefore:

$$(43.75)(1{,}728) = 75{,}600 \text{ cubic inches}$$

Cylindrical Containers

Cylindrical containers are found in the fire protection arena as water storage tanks on industrial property, storage tanks for oil and other combustible liquids, and sometimes forgotten, as containers for water on hose layouts. The formula commonly used in the fire service for determining the volume of a cylindrical container is:

$$V = .7854\,D^2 H$$

where   V = volume
        D = diameter
        H = height

Following is an example of the application of this formula.

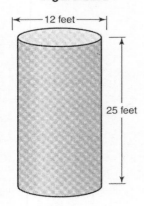

Figure 8.24

Problem: What is the volume of the cylindrical container shown in Figure 8.24?

Solution:

$$V = (.7854)(12)(12)(25)$$
$$= 2{,}827.44 \text{ cubic feet}$$

Problem: What would this equal in cubic meters?

Solution: To convert cubic feet to cubic meters, multiply by 0.03.

$$2827.44 \times 0.03 = 84.82 \text{ cubic meters}$$

Problem: What would be the volume of a rectangular tank that is 24 feet in height and has a diameter of 9 feet?

Solution:

$$V = .7854 \, D^2 H$$

where D = 9 feet
       H = 24 feet

then

$$V = (.7854)(9)(9)(24)$$
$$= 1{,}526.82 \text{ cubic feet (rounded off)}$$

Problem: What would this be equivalent to in cubic inches?

Solution: There are 1,728 cubic inches in a cubic feet.

Therefore, $(1{,}728)(1{,}526.82) = 2{,}638{,}344.96$ cubic inches.

Problem: What is the volume of the hose line shown in Figure 8.25?

Solution: A hose line is a cylindrical container. First, convert the length of the hose line to inches because all figures have to be in the same units: $100 \times 12 = 1{,}200$.

$$V = (.7854)(5)(5)(1{,}200)$$
$$= 23{,}562 \text{ cubic inches}$$

Problem: What would this equal in cubic milliliters?

Solution: To convert cubic inches to cubic milliliters, multiply by 16.39.

$$23{,}562 \times 16.39 = 38{,}6181.18 \text{ cubic milliliters}$$

Problem: What is the volume of a 100-feet length of 4-inch hose?

Solution: First, convert the length of the hose to inches by multiplying the length by 12.

$$100 \times 12 = 1{,}200$$
$$V = .7854 \, D^2 H$$

where D = 4 inches
       H = 1,200 inches

then

$$V = (.7854)(4)(4)(1{,}200)$$
$$= 15079.68 \text{ cubic inches}$$

Figure 8.25

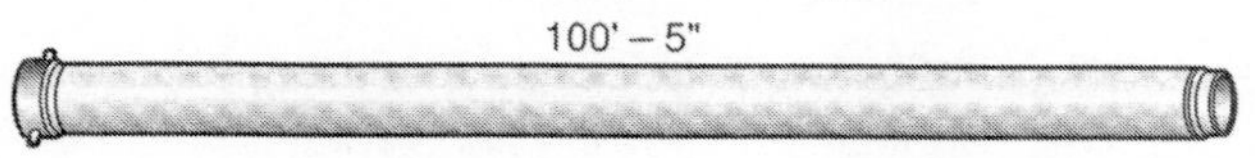

100' – 5"

Reader's Practice Problems—8.4

Feel free to use the conversion information introduced earlier in the chapter.

1. What is the volume in a water tank on an apparatus that measures 5.5 feet by 4 feet by 3 feet?
2. What is the volume equivalent to in cubic meters?
3. What is it equivalent to in cubic inches?
4. A small oil storage tank has dimensions of 4.2 feet by 2.4 feet by 2.2 feet. What is the volume capacity of the tank?
5. A cylindrical tank 28 feet in height has a 13-feet diameter. What is the volume capacity of this tank?
6. What would this be equivalent to in cubic meters?
7. What would this be equivalent to in cubic inches?
8. What is the volume of 350 feet of $3\frac{1}{2}$-inch hose?
9. What would this equal in cubic milliliters?
10. What is the volume of 500 feet of 2-inch hose?

The answers are found in the Reader's Practice Problems section of the Appendix.

VOLUME REVIEW QUESTIONS

1. Define volume.
2. How is the size of a volumetric container associated with the fire service normally expressed?
3. What type of container is normally used as a water tank on a fire apparatus?
4. What are the two most common types of containers used for storing water in the fire service?
5. What formula is used for determining the volume of a cylindrical container?
6. What formula is used for determining the volume of a rectangular container?
7. What formula is used for determining the volume of water in a hose line?
8. How many cubic inches are there in a cubic feet?
9. What is first necessary when determining the volume of a hose line?
10. In what units is the volume of a hose line normally given?
11. What is the standard unit for volume in the metric system?

Test 8.4

1. What is the volume of the hose line shown in Figure 8.26?
   SOLUTION HELP—V = .7854 D²H

   $$D = 4 \text{ inches}$$
   $$H = (50)(12)$$

Figure 8.26

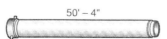

50' – 4"

Figure 8.27

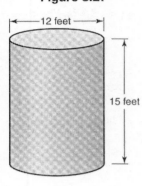

then
$$V = (.7854)(4)(4)(50)(12)$$
$$= 7{,}539.84 \text{ cubic inches}$$

2. What is the volume of the cylindrical container shown in Figure 8.27?

   SOLUTION HELP—$V = .7854\, D^2 H$

   where $D = 12$ feet

   $H = 15$ feet

   then
   $$V = (.7854)(12)(12)(15)$$
   $$= 1{,}696.46 \text{ cubic feet}$$

3. What is the volume of the rectangular container shown in Figure 8.28?

   SOLUTION HELP—$V = l\,w\,h$

   where $l = 15$ feet

   $w = 12$ feet

   $h = 20$ feet

   then
   $$V = (15)(12)(20)$$
   $$= 3{,}600 \text{ cubic feet}$$

4. How many cubic inches are there in 500 feet of 5-inch hose?

   SOLUTION HELP— $V = .7854\, D^2 H$

   where $D = 5$ inches

   $H = 500 \times 12$

   then
   $$V = (.7854)(5)(5)(500)(12)$$
   $$= 117{,}810 \text{ cubic inches}$$

5. What is the equivalent volume in cubic feet?

   SOLUTION HELP—Divide 117,810 by 1728.

Figure 8.28

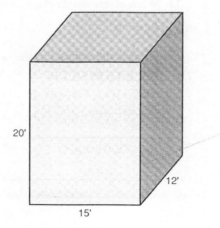

6. How many cubic feet are there in a rectangular container that measures 15 feet × 15 feet × 22 feet?

   SOLUTION HELP—V = l w h

   where   l = 22 feet
            w = 15 feet
            h = 15 feet

   then
   $$V = (22)(15)(15)$$
   $$= 4,950 \text{ cubic feet}$$

7. What would this be equivalent to in cubic inches?

   SOLUTION HELP—Multiply (4,950) (1,728)

8. A cylindrical container has a volume that measures 14,312 cubic feet. How tall is the container if it has a diameter of 10 feet?

   SOLUTION HELP—V = .7854 D²H

   V = 14,312

   then
   $$H = \frac{V}{(.7854)(D)(D)}$$
   $$H = \frac{14,312}{(.7854)(10)(10)}$$
   $$= 182.23 \text{ feet}$$

9. Two hose lines are taken off a pumper. One is 300 feet of 4-inch hose and the other is 500 feet of 3-inch hose. How many more cubic inches are in the hose having the largest volume than the one having the lesser volume?

   SOLUTION HELP—V = (.7854)(3)(3)(500)(12)   (3 inch)
   = 42,411.6 cubic inches

   V = (.7854)(4)(4)(300)(12)   (4 inch)
   = 45,239.04 cubic inches

   Subtract 42,411.6 from 45,239.04 = 2,827.44 cubic inches.

10. What is the volume of a rectangular container that measures 6 feet by 3 feet by 2 feet?

    SOLUTION HELP—V = l w h
    = 6 × 3 × 2
    = 36 cubic feet

11. Convert 500 cubic inches to milliliters.
12. Convert 500 cubic feet to cubic meters.
13. Convert 14 fluid ounces to milliliters.
14. Convert 7.2 gallons to liters.
15. Convert 850 milliliters to cubic inches.
16. Convert 7.8 cubic meters to cubic feet.
17. Convert 130 milliliters to fluid ounces.
18. Convert 9.5 liters to gallons.
19. At a small fire, Engine 2 has laid 550 feet of 2-inch hose while Engine 6 has laid 250 feet of $2\frac{1}{2}$-inch hose. How many cubic inches of hose has been laid by each company?

SOLUTION HELP—Use the formula .7854 D^2H to determine the cubic inches of hose laid. H equals the length of hose × 12.

$$\text{Engine 2's line} - .7854 \times 2 \times 2 \times 550 \times 12$$
$$\text{Engine 6's line} - .7854 \times 2.5 \times 2.5 \times 250 \times 12$$

20. The rectangular water tank on Engine 5 measures 5 feet by 4 feet by 2.5 feet in height. How many cubic inches of volume are there in this container?

    SOLUTION HELP—The formula to use is:

    $$V = l \times w \times h \times 1{,}728$$

    Multiply $5 \times 4 \times 2.5 \times 1{,}728$.

21. What is the volume in a cylindrical container that has a radius of 10 feet and is 25 feet high?

    SOLUTION HELP—The diameter is 20 feet (2 × radius).

    The formula to use is: $V = .7854\, D^2H$

    Multiply $.7854 \times 20 \times 20 \times 25$.

22. What would the volume be in cubic inches?

    SOLUTION HELP—Multiply 7854×1728.

23. What would the volume be in 650 feet of 4-inch hose?

    SOLUTION HELP—Use the formula:

    $$V = .7854\, D^2H$$

    where   H = length of hose × 12

    Multiply $.7854 \times 4 \times 4 \times 650 \times 12$.

24. Are there more cubic inches in 250 feet of 5-inch hose? If so, how many?

    SOLUTION HELP—Use the formula:

    $$V = .7854\, D^2H$$

    where   H = length of hose × 12

    Multiply $.7854 \times 5 \times 5 \times 250 \times 12$.

25. What is the volume in a rectangular container that measures 12 feet by 12 feet by 8 feet in height?

    SOLUTION HELP— Use the formula:

    $$V = l\, w\, h$$

    Multiply $12 \times 12 \times 8$.

Weight Measurements

Weight is the downward force created by an object due to the earth's gravity. In the customary system, the standard unit of measurement is the ***pound***. In the fire service, the weight of water is used as a standard. The weight of a column of water one feet high with a base of one square inch weighs approximately .434 pounds. A column of water measuring 1 inch by 1 inch at the base that is 2.304 feet high weighs approximately 1 pound. A gallon of water weighs approximately 8.35 pounds. There are approximately 62.5 pounds of water in a cubic feet.

The standard unit of mass (weight) in the metric system is the gram. The other units and their equivalents in grams are:

$$1 \text{ milligram (mg)} = 0.001 \text{ gram}$$
$$1 \text{ centigram (cg)} = 0.01 \text{ gram}$$
$$1 \text{ decigram (dg)} = 0.1 \text{ gram}$$
$$1 \text{ kilogram (kg)} = 1000 \text{ grams}$$

The approximate conversion of the weight of water from the customary system to the metric system can be accomplished by multiplying the weight of the water in pounds by 0.45 to obtain the equivalent number of kilograms. Consequently, to determine the weight of one gallon of water in the metric system, multiply 8.35 by 0.45.

$$8.35 \times 0.45 = 3.76 \text{ kilograms}$$

To determine the weight of one cubic feet of water in the metric system, multiply 62.5 by 0.45.

$$62.5 \times 0.45 = 28.13 \text{ kilograms}$$

Determining the Weight of Water in a Container

The formula for determining the weight of water in a rectangular or cylindrical container is:

$$\text{weight} = 62.5 \text{ V}$$

Figure 8.29

Problem: What would be the weight of the water in the rectangular container shown in Figure 8.29 when the container is full?

Solution:

$$\begin{aligned}\text{weight} &= 62.5 \text{ V}\\ &= (62.5)(54)\\ &= 3{,}375 \text{ pounds}\end{aligned}$$

Problem: What would this be equivalent to in kilograms?

Solution: To change the weight of the water to kilograms, multiply the weight in kilograms of one cubic feet of water (28.13) by the number of cubic feet in the container.

$$= 1519.02$$

Problem: A cylindrical tank is 30 feet high and has a diameter of 16 feet. If this tank is full of water, how much would the water weigh?

Solution: It is first necessary to determine the volume of the container.

$$V = .7854 \, D^2 H$$

where $D = 16$ feet
$H = 30$ feet

then

$$\begin{aligned}V &= (.7854)(16)(16)(30)\\ &= 6{,}031.87 \text{ cubic feet (rounded off)}\end{aligned}$$

$$\begin{aligned}\text{weight} &= 62.5 \text{ V}\\ &= (62.5)(6{,}031.87)\\ &= 3{,}76{,}991.88 \text{ pounds (rounded off)}\end{aligned}$$

Figure 8.30

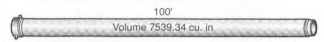

Problem: What would be the weight of the water in the hose line shown in Figure 8.30 when the line is full?

Solution: The determination of the weight of water in a hose line is a two-step process. It is first necessary to determine the number of gallons of water in the line. The gallon capacity of the hose line can be determined by the formula:

$$\text{gallon capacity} = \frac{\text{volume (cu. in.)}}{231}$$

It was determined that the volume of the hose line shown in Figure 8.30 was 7,539.34 cubic inches. Therefore, the gallon capacity is:

$$\text{gallons} = \frac{7{,}539.34}{231} = 32.64 \text{ gallons}$$

The formula to change gallons to weight is:

$$\begin{aligned}\text{weight} &= \text{gallons} \times 8.35 \\ &= (32.64)(8.35) \\ &= 272.54 \text{ pounds}\end{aligned}$$

Problem: How much does the water weigh in a 50-feet section of $2\frac{1}{2}$-inch hose?

Solution: First, determine the number of gallons of water in the hose line.

$$\text{gallon capacity} = \frac{.7854\, D^2 H}{231}$$

where $D = 2.5$ inches
$H = 50 \text{ feet} \times 12 \text{ inches} = 600 \text{ inches}$

then

$$\begin{aligned}\text{gallon capacity} &= \frac{(.7854)(2.5)(2.5)(600)}{231} \\ &= \frac{2{,}945.25}{231} \\ &= 12.75 \text{ gallons} \\ \text{weight} &= 8.35 \times \text{gallons} \\ &= (8.35)(12.75) \\ &= 106.46 \text{ pounds (rounded off)}\end{aligned}$$

Reader's Practice Problems—8.5

1. A rectangular container has a length of 8 feet, a width of 6 feet, and a height of 4 feet. If it is filled with water, how much will the water weigh?
2. How much will the water weigh in a 400-feet length of 4-inch hose?
3. Two five-gallon jugs are each filled with water. How much will the water weigh in the two jugs?

4. The storage capacity of a cylindrical container is 5,784 cubic feet. If filled with water, how much will the water weigh?
5. A 22-feet high cylindrical container has a diameter of 12 feet. If filled with water, how much will the water weigh?
6. How much will the water weigh in a 100-feet length of 3-inch hose?

The answers are found in the Reader's Practice Problems section of the Appendix.

WEIGHT REVIEW QUESTIONS

1. Define weight.
2. What is the standard unit for weight in the customary system?
3. How much will a column of water weigh that is one feet high and has a base of one square inch?
4. How high is a column of water that has a base of one square inch and weighs one pound?
5. How much does a gallon of water weigh?
6. How many pounds of water are there in a cubic feet?
7. What is the formula for determining the weight of water in a rectangular or cylindrical container?
8. What is the two-step process for determining the weight of water in a hose line?
9. What is the formula for determining the number of gallons of water in a hose line?
10. What is the formula for determining the weight of water in a hose line?
11. How do you change the weight of a gallon of water to kilograms?
12. How do you change the weight of a cubic feet of water to kilograms?
13. What is the standard unit of mass (weight) in the metric system?

Test 8.5

1. How much will the water weigh in a full rectangular container that has a volume of 62.4 cubic feet?
2. How much will the water weigh in a full cylindrical container that has a volume of 91.3 cubic feet?
3. How many gallons of water will a 50-feet section of 3 inch hose hold?
4. How much does the water weigh in a 50-feet section of 3-inch hose?
5. How much greater is the weight of water in a full rectangular container that has a volume of 72.6 cubic feet than in a full cylindrical container that has a volume of 41.4 cubic feet?
6. What would be the weight of water in a full rectangular container that measures 8 feet by 10 feet by 16 feet?

SOLUTION HELP—weight = 62.5 V

$$\text{volume} = l\ w\ h$$

First, find the volume and then multiply the volume by 62.5.

$$\text{volume} = 8 \times 10 \times 16$$

7. There are 7.48 gallons in a cubic feet. How many gallons can the rectangular container in Problem 6 hold?

   SOLUTION HELP— Use the formula:
   $$\text{Gallons} = 7.48\,V$$
   $$V = 8 \times 10 \times 16$$
   Then $\text{gallons} = 7.48 \times 8 \times 10 \times 16$

8. Using the information in Problems 6 and 7, how much does a gallon of water weigh?

   SOLUTION HELP—
   $$\text{weight of one gallon of water} = \frac{\text{weight of water in container}}{\text{gallons in container}}$$
   $$\text{Divide} \quad \frac{80{,}000}{9574.4}$$

9. What would the water weigh in a full cylindrical container that that has a diameter of 15 feet and is 30 feet in height?

   SOLUTION HELP—Use the formula $62.5\,V$, wherein:
   $$V = .7854\,D^2 H$$
   Multiply $62.5 \times .7854 \times 15 \times 15 \times 30$.

10. How many gallons will the container in Problem 9 hold?

    SOLUTION HELP—Use the formula:
    $$\text{gallons} = 7.48\,V$$
    Multiply $7.48 \times .7854 \times 15 \times 15 \times 30$.

Fluid Measurements

Fluid is a substance that flows freely. It has no fixed shape and little resistance to outside stress. The fluid that will be considered in this section is water. Water in fire protection is generally stored in either a rectangular container or a cylindrical container and forced to flow through hose lines. The amount of water stored in a rectangular or cylindrical container can be determined by the formula:

$$\text{gallon capacity} = 7.48\,V$$

where V = volume in cubic feet

An example of the use of this formula is illustrated in the following problem.

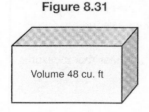

Figure 8.31

Problem: How many gallons of water can be stored in the container shown in Figure 8.31?

Solution:
$$\text{gallon capacity} = 7.48\,V$$
$$= (7.48)(48)$$
$$= 359.04 \text{ gallons}$$

Problem: A rectangular container measures 7.5 feet by 5 feet by 3.5 feet. If filled with water, how many gallons will it hold?

Solution: First, determine the volume of the container.
$$V = l\,w\,h$$

where   l = 7.5 feet
         w = 5 feet
         h = 3.5 feet

then
$$V = (7.5)(5)(3.5)$$
$$= 131.25 \text{ cubic feet}$$
$$\text{gallon capacity} = 7.48V$$
$$= (7.48)(131.25)$$
$$= 981.75 \text{ gallons}$$

Problem: A cylindrical tank having a diameter of 16 feet is 28 feet tall. When full, how many gallons of water will it hold?

Solution: First, determine the volume of the container.

$$V = .7854 \, D^2H$$

where   D = 16 feet
         H = 28 feet

then
$$V = (.7854)(16)(16)(28)$$
$$= 5{,}629.75 \text{ cubic feet} \quad \text{(rounded off)}$$

then
$$\text{gallon capacity} = 7.48V$$
$$= (7.48)(5629.75)$$
$$= 42{,}110.53 \text{ gallons}$$

Although a hose line is actually a cylindrical container, the formula used to determine the number of gallons of water in a hose line is different than the one used for determining the gallons of water in a standard storage container. The formula used is:

$$\text{gallon capacity} = \frac{\text{volume (in cubic inches)}}{231}$$

An example of its use is illustrated in the following problem.

Problem: How many gallons of water will the hose line shown in Figure 8.32 hold?

Solution:

$$\text{gallons capacity} = \frac{\text{volume (cu. in.)}}{231}$$
$$= \frac{2{,}500}{231}$$
$$= 10.82 \text{ gallons}$$

Problem: How many gallons of water will a 350-feet length of 3-inch hose hold?

Solution: First, determine the volume of the hose line.

$$V = .7854 \, D^2H$$

Figure 8.32

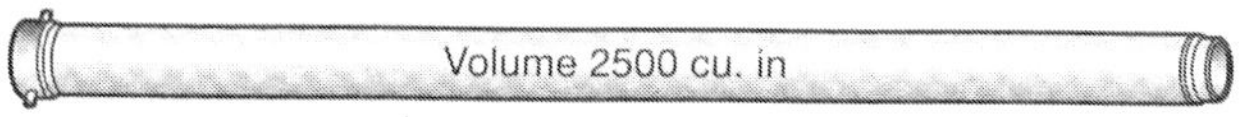

Volume 2500 cu. in

where   D = 3 inches
H = 350 feet × 12 inches = 4,200 inches

then
$$V = (.7854)(3)(3)(4,200)$$
$$= 29,688.12 \text{ cubic inches}$$

then
$$\text{gallons capacity} = \frac{\text{volume (cu. in.)}}{231}$$
$$= \frac{29,688.12}{231}$$
$$= 128.52 \text{ gallons}$$

Reader's Practice Problems—8.6

1. How much water will a rectangular tank hold that measures 12 feet long by 8 feet wide by 5 feet in height?
2. A cylindrical tank is 65 feet tall and has a diameter of 23 feet. How much water will this tank hold when full?
3. How much water will a 750-feet length of $1\frac{1}{2}$-inch hose hold?
4. How much water will a 650-feet length of 4-inch hose hold?

The answers are found in the Reader's Practice Problems section of the Appendix.

FLUID REVIEW QUESTIONS

1. Define fluid.
2. What formula is used for determining the number of gallons of water in a rectangular container?
3. What formula is used for determining the number of gallons of water in a cylindrical container?
4. What formula is used for determining the number of gallons of water in a hose line?
5. What must be done to the length of a hose line to determine the number of gallons of water that the hose line can hold when using the hose line formula?

Test 8.6

1. The firefighting helicopter (Figure 8.33) picking up water can fill its water tank in just a few minutes. The tank has a volume capacity of 180 cubic feet. How many gallons of water will the helicopter pick up on each load?

   SOLUTION HELP—gallons = 7.48 V
   $$= (7.48)(180)$$
   $$= 1346.4$$

Figure 8.33

Courtesy of Rick McClure, LAFD

2. How many gallons of water will the hose line shown in Figure 8.34 hold?

   SOLUTION HELP—gallons = 7.48 V

   $$V = 28{,}628 \text{ cubic inches} \div 1{,}728$$
   $$= 16.57 \text{ cubic feet}$$

   then
   $$\text{gallons} = (7.48)(16.57)$$
   $$= 123.9 \text{ gallons}$$

3. How many gallons of water will a rectangular container hold if it has a volume of 29.7 cubic feet?

   SOLUTION HELP—gallons = 7.48 V
   $$= (7.48)(29.7)$$
   $$= 222.16 \text{ gallons}$$

Figure 8.34

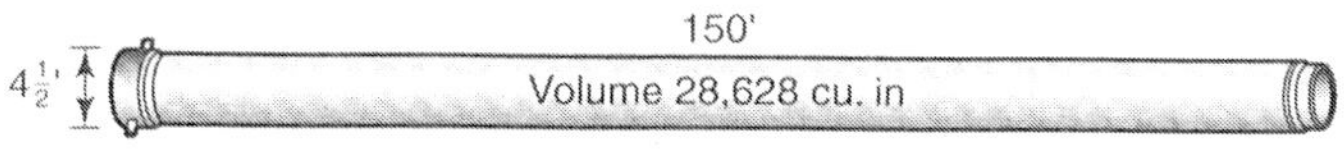

$4\frac{1}{2}'$ 150' Volume 28,628 cu. in

4. How many gallons of water will a cylindrical container hold if it has a volume of 42.4 cubic feet?

   SOLUTION HELP—gallons = 7.48 V
   $$= (7.48)(42.4)$$
   $$= 317.15 \text{ gallons}$$

5. A rectangular container has a length of 5 feet, a width of 3 feet, and a height of 4 feet. Compared to a 50-feet section of 6-inch hose, does the container hold more water or does the hose hold more water? What is the difference in volume between the hose and the container?

   SOLUTION HELP—gallons = 7.48 V
   $$V = l\,w\,h$$
   $$\text{gallons} = (7.48)(5)(3)(4) \quad \text{(rectangle)}$$
   $$= 448.8$$

   $$\text{gallons} = \frac{V}{231}$$
   $$V = (.7854)(6)(6)(50)(12)$$
   $$\text{gallons} = \frac{(.7854)(6)(6)(50)(12)}{231}$$
   $$= \frac{16{,}964.64}{231}$$
   $$= 73.44 \text{ gallons} \quad \text{(hose)}$$

   Subtract 73.44 from 448.8 = 375.36 gallons (rectangular)

6. A helicopter that has a water tank with a volume of 200 cubic feet is picking up its second load for the day. How many gallons of water can it pick up?

   SOLUTION HELP— Use the formula:

   $$\text{gallons} = 7.48\,V$$

   Multiply 7.48 × 200.

7. How many gallons of water will a 450-feet line of 4-inch hose hold?

   SOLUTION HELP— Use the formula:

   $$\text{gallons} = \frac{\text{volume (cubic inches)}}{231}$$

   where   volume = .7854 D²H
   $$H = \text{hose length} \times 12$$

   $$\text{volume} = \frac{(.7854)(4)(4)(450)(12)}{231}$$

8. A cylindrical storage tank has a 24-feet diameter and is 44 feet high. How many gallons of water can it hold?

   SOLUTION HELP—Use the formula:

   $$\text{gallons} = 7.48\,V$$

where $V = .7854 D^2 H$

$V = (.7854)(24)(24)(44)$

Multiply $7.48 \times .7854 \times 24 \times 24 \times 44$.

9. A rectangular tank measures $6 \times 8 \times 12$ inches. How many gallons of water can it hold?

SOLUTION HELP—Use the formula:

$$gallons = 7.48 V$$

where $V = l\,w\,h$

$V = (6)(8)(12)$

Multiply $7.48 \times 6 \times 8 \times 12$.

10. Which holds more gallons, the tank in Problem 9 or 500 feet of 5-inch hose?

SOLUTION HELP—Use the formula:

$$gallons = \frac{volume}{231}$$

where $volume = .7854 D^2 H$

$H = 500 \times 12$

then $$gallons = \frac{(.7854)(5)(5)(500)(12)}{231}$$

Temperature Measurements

Temperature is a degree of heat. It is expressed as hot or cold relative to something else. For example, ***body heat*** is the degree of heat in a living organism. For fire and EMS personnel, the living organism is a human being. In the United States, the standard body temperature is 98.6°F. In almost all other parts of the world, it is 37°C. A human being with a body heat above these standards is considered to have a ***fever***. An example of the comparison between the two systems is shown in Figure 8.35.

It should be noted that with the ***centigrade*** (or Celsius) system that water freezes at zero degrees and boils at 100 degrees. This appears to be more practical than the ***Fahrenheit*** system, which has a spread of 180 degrees between freezing and boiling water. It should be further noted that zero in either of the two systems does not refer to absolute zero. Absolute zero refers to a temperature at which all oscillations of a body's atoms and molecules theoretically stop. Absolute zero occurs in a Fahrenheit system at −459 degrees. It occurs in the Centigrade system at −273 degrees.

Two formulas are associated with the conversion of one system to another. To convert Fahrenheit temperature to Centigrade temperature, the following formula is used:

$$C = \frac{(5)(F - 32)}{9}$$

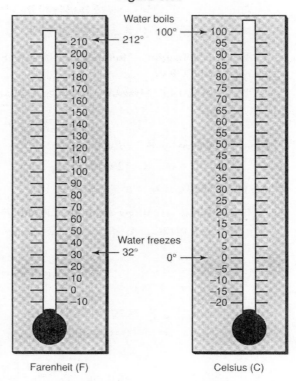

Figure 8.35

Problem: Convert 80 degrees Fahrenheit to degrees Centigrade (Figure 8.35).

Solution:

$$C = \frac{(5)(80-32)}{9}$$

$$= \frac{(5)(48)}{9}$$

$$= \frac{240}{9}$$

$$= 26.67$$

The formula for converting Centigrade temperature to Fahrenheit temperature is as follows:

$$F = \frac{9C}{5} + 32$$

Problem: Convert 20 degrees Centigrade to degrees Fahrenheit.

Solution:

$$F = \frac{(9)(20)}{5} + 32$$

$$= \frac{180}{5} + 32$$

$$= 36 + 32$$

$$= 68 \text{ degrees}$$

Firefighters are subjected to a greater change in atmospheric conditions than those working in any other occupation. Despite having to work in atmospheres where the temperatures change rapidly, there are no instruments available that normally record these rapidly changing temperatures during firefighting operations. Therefore, no information is available for inclusion in this chapter regarding measuring temperature changes during firefighting operations.

TEMPERATURE REVIEW QUESTIONS

1. Define temperature.
2. Define body heat.
3. Define fever.
4. What is the standard body temperature in Fahrenheit?
5. What is the standard body temperature in Centigrade?
6. What formula is used for converting Fahrenheit temperature to Centigrade temperature?
7. What formula is used for converting Centigrade temperature to Fahrenheit temperature?
8. What is the spread between where water freezes and water boils in the Centigrade system?
9. What is the spread between where water freezes and water boils in the Fahrenheit system?
10. A material has an ignition temperature of 840 degrees Fahrenheit. What would be its ignition temperature in Centigrade?

Test 8.7

1. Convert 100 degrees Fahrenheit to degrees Centigrade.
2. Convert 36 degrees Fahrenheit to degrees Centigrade.
3. Convert 48 degrees Fahrenheit to degrees Centigrade.
4. Convert 30 degrees Centigrade to degrees Fahrenheit.
5. Convert 48 degrees Centigrade to degrees Fahrenheit.
6. When first taking the vital signs of a patient, a paramedic gets a temperature reading of 98.8. Forty minutes later she gets a reading of 103.2. How much has the patient's temperature risen?
7. When first taking the vital signs of a patient, a paramedic gets a temperature reading of 104.7. After following the proper procedure for lowering temperature, the reading is 99.9. How much improvement has been achieved by following the proper procedure?
8. The flashpoint of a flammable liquid is 4 degrees Fahrenheit. What would its flashpoint be in Centigrade?
9. The ignition temperature of a material is 850 degrees Centigrade. What would its ignition temperature be in Fahrenheit?
10. The ceiling temperature of a room when the firefighters first entered was 525 degrees Fahrenheit. What would this be in Centigrade?

Chapter Summary

This chapter introduced the measurement methods used in the fire service and Emergency Medical Service, and was divided into separate sections on length, area, pressure, volume, weight, fluid, and temperature.

Each section included the definition of words used in that measurement system, formulas used with the system, problems to illustrate the use of the formulas, and problems for the reader to solve. The end of each section then included applicable review questions.

In addition, the chapter provided information and charts for determining the equivalencies between the units of measurement in the metric system and the units of measurement in the customary system.

CHAPTER 9
WORKING WITH PERCENT AND PERCENTAGES

KEY TERMS

decimal	percentage	ratio
fraction	percentile	sale items
percent	per centum	tipping

OBJECTIVES AND GOALS

Upon completing this chapter, the reader should be able to:

- Define percent and percentage.
- Define ratio.
- Define percentile.
- Convert a percent to a decimal.
- Convert a decimal to a fraction.
- Convert a percent to a fraction.
- Convert a decimal to a percent.
- Convert a fraction to a percent.
- Solve problems associated with percentages.
- Solve percent problems.
- Solve alternative percent problems.
- Explain how to use the $3 \times 4 = 12$ principle.
- Explain the principles of tipping.
- Discuss the merits and hazards of sale items.

WORDS TO REMEMBER

ratio—A term used to compare two quantities.

percent—A portion of a whole. The whole is one hundred. A percent has an identification symbol (%) attached to the end.

percentage—Also a portion of a whole but does not have an identification symbol.

Introduction to the Term Ratio

Ratio is a term used to compare two numbers such as 7 and 8 or two of the same type of measurement such as 250 gpm (gallons per minute) to 315 gpm. A ratio may be expressed as 7 to 8, 7 ÷ 8, or $\frac{7}{8}$. When written as 7 ÷ 8, the symbol ÷ indicates "to" 8. When written as a fraction, the line dividing the numerator from the denominator is also referred to as "to."

Introduction to Percent

Like decimals and fractions, percent is a portion of a whole. However, whereas the basic whole for decimals and fractions is one, the whole for percent is one hundred. In fact, the word itself is derived from the Latin words *per centum*, which means "by a hundred."

Percent also has the honor of having its own identification symbol (%). In fact, the symbol shares a numerical key (5) on the standard computer keyboard. The symbol is almost always included when referring to mortgage rates, cost of living, and other factors such as sales promotions. Occasionally, it is advantageous to convert a percent to a decimal fraction or a common fraction. At other times it is useful to convert a fraction or a decimal fraction to a percent. Following are suggestions for making these conversions.

Converting a Percent to a Decimal

To convert a percent to a *decimal*, simply drop the percent symbol and move the decimal point two places to the left. In making this conversion, it should be remembered that a percent figure is a whole number and all whole numbers have an unwritten decimal point to the right. Following are several examples of the proper procedure.

Problem: Convert 66% to a decimal.

Solution:

   Step 1. Drop the % symbol: 66% becomes 66.

   Step 2. Move the decimal point two places to the left: 66 becomes .66.

Problem: Convert 72.5% to a decimal.

Solution:

   Step 1. Drop the % symbol: 72.5% becomes 72.5.

   Step 2. Move the decimal point two places to the left: 72.5 becomes .725.

Converting a Decimal to a Fraction

A decimal can be converted to a *fraction* by first placing the decimal over 1. Then repeatedly multiply both the numerator and the denominator by 10 until the numerator is a whole number.

Problem: Convert .64 to a fraction.

Solution:

   Step 1. Place .64 over 1: $\frac{.64}{1}$

   Step 2. Multiply the numerator and denominator by 10.

$$\frac{.64}{1} \times \frac{10}{10} = \frac{6.4}{10}$$

Step 3. Multiply the numerator and denominator by 10.

$$\frac{6.4 \times 10}{10 \times 10} = \frac{64}{100}$$

Step 4. Change the numerator and denominator of the fraction into their respective prime numbers.

$$\frac{64}{100} = \frac{(2)(2)(2)(2)(2)(2)}{(2)(2)(5)(5)}$$

Step 5. Cancel the first two 2's of both the numerator and denominator.

$$\frac{64}{100} = \frac{(2)(2)(2)(2)(2)(2)}{(2)(2)(5)(5)}$$

Step 6. Multiply the prime numbers in both the numerator and denominator to obtain a fraction reduced to its lowest terms.

$$\frac{(2)(2)(2)(2)}{(5)(5)} = \frac{16}{25}$$

Converting a Percent to a Fraction

Converting a percent to a fraction is a two-step process. First, convert the percent to a decimal by dropping the % symbol and moving the decimal point two places to the left. Next, convert the decimal to a fraction. (This procedure was explained in a previous chapter, however, it is repeated here to save the reader the necessity of flipping back through the pages.)

Problem: Convert 68% to a fraction.

Solution:

Step 1. Convert 68% to a decimal. Drop the percent symbol %: 68% becomes 68.

Step 2. Move the decimal point two places to the left: 68 becomes .68.

Step 3. Use the procedure provided in the previous section to change the decimal to a fraction.

Converting a Decimal to a Percent

It is a simple process to change a decimal to a percent. Simply move the decimal point two places to the right and add the percent symbol.

Problem: Convert the decimal .046 to a percent.

Solution:

Step 1. Move the decimal point in the decimal .046 two places to the right: .046 becomes 4.6.

Step 2. Add the percent symbol: 4.6 becomes 4.6%.

Problem: Convert the decimal 2.43 to a percent.

Solution:

Step 1. Move the decimal point in the decimal 2.43 two places to the right: 2.43 becomes 243.

Step 2. Add the percent symbol: 243 becomes 243%.

Converting a Fraction to a Percent

Converting a fraction to a percent is a three-step process.

Step 1. Divide the numerator by the denominator.

Step 2. Move the decimal point obtained two places to the right.

Step 3. Add the percent symbol.

Problem: Convert $\frac{3}{7}$ to a percent.

Solution:

Step 1. Divide the numerator (3) by the denominator (7).

$$7\overline{)3.0} = .429$$

Step 2. Move the decimal point in the answer two places to the right.

.429 becomes 42.9

Step 3. Add the percent symbol.

42.9 becomes 42.9%

Problem: Convert $\frac{3}{4}$ to a percent.

Solution

Step 1. Divide the numerator (3) by the denominator (4).

$$4\overline{)3.0} = .75$$

Step 2. Move the decimal point in the answer two places to the right.

.75 becomes 75

Step 3. Add the percent symbol.

75 becomes 75%

If the fraction is a mixed number, first convert the mixed number to an improper fraction and proceed as shown above.

Problem: Convert $3\frac{3}{8}$ to a percent.

Solution:

Step 1. Convert $3\frac{3}{8}$ to an improper fraction. It becomes:

$$\frac{27}{8}$$

Step 2. Divide the numerator by the denominator.

$$8\overline{)27.000} = 3.375$$

Step 3. Move the decimal point of the answer two places to the right.

3.375 becomes 337.5

Step 4. Change the answer to a percent by adding the percent symbol.

337.5 becomes 337.5%

Problem: Convert $13\frac{3}{7}$ to a percent.

Solution:

Step 1. Convert $13\frac{3}{7}$ to an improper fraction. It becomes:

$$\frac{94}{7}$$

Step 2. Divide the numerator by the denominator.

$$7\overline{)94.0000} = 13.4286$$

Step 3. Move the decimal point of the answer two places to the right.

13.4286 becomes 1342.86

Step 4. Change the answer to a percent by adding the percent symbol.

1342.86 becomes 1342.86%

Take a break and solve the following problems. Use the samples just provided to determine the procedure to use. If you have any trouble working any of the problems, refer back and review the samples that were provided.

Problem 1. Convert 57% to a decimal.

Problem 2. Convert 88% to a fraction. Reduce the fraction to its lowest terms.

Problem 3. Convert the decimal .64 to a percent.

Problem 4. Convert the fraction $\frac{7}{10}$ to a percent.

Problem 5. Convert .35 to a fraction.

Solving Percent Problems

Solving a percent problem is a simple procedure of changing the percent to a decimal and proceeding as if the problem is a multiplication problem involving decimals. As a review, the number being multiplied is referred to as the multiplicand, the number that is doing the multiplying (in this situation, it is the percent changed to a decimal) is called the multiplier, and the answer is referred to as the product. Using the problem, "What is 12% of 234?" as an example, the three portions of the problem would appear as:

$$\begin{array}{r} 234 \\ \times\ .12 \\ \hline 28.08 \end{array}$$ multiplicand (the number being multiplied)
multiplier (the percent changed to a decimal)
product (answer to the multiplication)

Another example:

Problem: What is 17% of 193?

Solution:

Step 1. Change the 17% to a decimal by dropping the percent symbol and moving the decimal point two places to the left: 17% becomes .17.

Step 2. Set up the multiplication problem by making the 193 the multiplicand and the percent changed to a decimal the multiplier.

$$\begin{array}{r} 193 \\ \times\ .17 \end{array}$$

Step 3. Complete the multiplication.

$$\begin{array}{r} 193 \\ \times\ .17 \\ \hline 32.81 \end{array}$$

Problem: The sales tax on groceries in city X is 8.3%. The total for the groceries that Mr. Smith bought was $52.70. How much was the sales tax?

Solution:

Step 1. Change the percent of the sales tax to a decimal by dropping the percent symbol and moving the decimal point two places to the left: 8.3% becomes .083.

Step 2. Set up the multiplication problem with the total cost of the groceries as the multiplicand and the percent changed to a decimal as the multiplier.

$$\begin{array}{r} 52.70 \\ \times\ .083 \\ \hline \$4.37 \end{array}$$

Problem: The salary of an EMT for an ambulance company was $1,682 a month. During the annual salary negotiations she was granted a 4.7% raise in pay. What is her new monthly salary?

Solution:

Step 1. Change the 4.7% raise to a decimal by dropping the percent symbol and moving the decimal point two places to the left: 4.7% becomes .047.

Step 2. Multiply her present salary (1,682) by the percent changed to a decimal (.047) to determine the increase in pay she received.

$$\begin{array}{r} 1682 \\ \times\ .047 \\ \hline \$79.05 \end{array}$$

Step 3. Add the amount of the raise to her present salary to obtain her new salary.

$$\$1{,}682 + \$79.05 = \$1{,}761.05$$

Solving Alternative Percent Problems

At times both the multiplier (percent) and the product (answer) are known and it is desired to determine the multiplicand (the base that is being multiplied). This type of problem is confusing to many individuals. As an example, a problem may ask, "77 is 9% of what number?" Rather than providing a formula for this type of problem and additional formulas for other types of alternative percent problems for the reader to memorize, a procedure is offered than can be used for these and other types of problems. The procedure is referred to as the 3 × 4 = 12 concept.

THE 3 × 4 = 12 CONCEPT

The problem used to explain how to solve percent problems is used to explain this concept. Consider that the 4 is the multiplicand in a problem, the 3 is the multiplier, and the 12 is the product. The problem would then appear as:

$$\begin{array}{c} 4 \\ \times\ 3 \\ \hline 12 \end{array} \quad \text{think of it as} \quad \begin{array}{c} \text{number} \\ \times\ \text{percent (expressed as a decimal)} \\ \hline 12 \end{array} \text{result}$$

Now use the problem "77 is 9% of what number?" to demonstrate the principle.

The number is missing	X
The 3 is the percent (9%)	.09
The 12 is the result (77)	77

In the $3 \times 4 = 12$ problem, if the 4 were missing, you would divide the 12 by 3 to find it. Using this thinking in the given problem, to determine the X you would need to divide the 77 by .09.

$$77 \text{ divided by } 0.09 = 855.56$$

Therefore, 9% of 855.56 = 77.

To test the validity of the problem, solve the problem "determine 9% of 855.56."

Solution:

Step 1. Change the 9% to a decimal by dropping the percent symbol and moving the decimal point two places to the left. The result is .09.

Step 2. Multiply the 855.56 by the .09.

$$\begin{array}{r} 855.56 \\ \times\ .09 \\ \hline 77 \end{array}$$

Following is an example of a different type of problem using the $3 \times 4 = 12$ concept to solve it.

Problem: What percent of 425 equals 8.25?

Solution:
The number 4 is the	425
The number 3 is missing	X
The number 12 is the	8.25

If the 3 is missing in the $3 \times 4 = 12$ problem, it would be necessary to divide the 12 by the 4 to determine it. In this problem it is necessary to divide the 8.25 by the 425 to determine the percent: 8.25 divided by 425 = 0.0194. To change it to a percent, move the decimal two places to the right and add the percent symbol: 0.0194 becomes 1.94%. Therefore, 1.94% of 425 = 8.25.

Following is another example of a practical problem.

Problem: A $1\frac{1}{4}$-inch tip on a $2\frac{1}{2}$-inch line is capable of effectively discharging approximately 326 gpm. Due to a reduction of the rpm of the apparatus as the result of a malfunction, the tip is discharging only 185 gpm. What percent of its capability is the tip discharging?

Solution: Using the $3 \times 4 = 12$ concept,

The 326 is the	4
The % missing	X
The 185 is the	12

To determine the X in the $3 \times 4 = 12$ problem, it is necessary to divide the 12 by the 4. In the given problem, to determine the percent it is necessary to divide the 185 by the 326.

$$185 \text{ divided by } 326 = 0.5675$$

To change this to a percent, move the decimal point two places to the right and add the percent symbol: 0.5675 becomes 56.75%. Therefore, the tip is discharging 56.75% of its capability.

Introduction to Percentage

Percentage is also part of a whole; however, it differs from percent in the fact that it does not have a symbol attached. Percentage problems can be solved by the formula:

$$P = R \times B$$

where P = percentage
R = rate (the percent expressed as a decimal)
B = base (the whole to which the percentage is a part)

Problem: What percentage is 18% of 250?

Solution: First, convert the 18% to a decimal: 18% = .18.

Then, P = .18 × 250 = 45 (this means that 18% of 50 is 45).

It should be noted that the above explanation of how to determine percentage is the same as the previous explanation of how to determine the percent of a number. The change is that a formula and letters have been introduced to the procedure. The 3 × 4 = 12 concept can be used with the letters placed in their proper position.

Introduction to Percentile

Many individuals confuse the term percentile with percent. In fact, some individuals think that a student who is placed in the 88 percentile of an examination he has taken has received a grade of 88% on the examination. This is false thinking. A *percentile* is a value based on a scale of one hundred that indicates the percent of a distribution that is equal to or below it. In other words, a student who is placed in the 88 percentile of an examination has received a grade that is equal to or better than 88% of the students who took the examination. The normal of the students who took the examination has a percentile between 25 and 75. A percentile greater than 75 is considered above normal whereas a percentile of less than 25 is considered to be below normal.

Tipping

Tipping is a standard part of "eating out." Almost all individuals tip whenever they have a meal in a restaurant. The difference is that different individuals tip different amounts. It appears that the three most common amounts are 10%, 15%, and 20%.

Regardless of which of these is used by an individual, unless an individual carries a calculator, it is necessary to determine the amount to tip mentally. This requirement is very difficult for some individuals. However, if the following procedures are used, the problem becomes very simple.

First of all, it is necessary to determine the base to use for figuring the tip. Some individuals tip on the cost of the meal only. Others tip on the total bill, which includes the cost of the meal and the tax. Most authorities state that tipping on the cost of the meal only is entirely acceptable. Regardless, once the base is established, one of the following procedures can be used to determine the amount to tip.

If the standard used by an individual is 10%, simply move the decimal point one place to the left. For example, if the cost of the meal is $18.67 and the cost of the meal is used as the base, moving the decimal point one place to the left will result in a tip of $1.87. If the base for the meal is $43.54, the result would be $4.35.

If the standard used by an individual is 15%, then an additional two steps are required. After determining the amount based on 10%, divide the amount by 2 and add the result to the 10% amount. For example, if the amount determined by 10% is $1.80, dividing the $1.80 by 2 = .90. Adding this to the $1.80 results in an amount of $2.70. The $2.70 is 15% of the base.

Determining a tip of 20% is even easier than determining a 15% tip. After using the system of 10%, simply double this amount. For example, if the tip on 10% would be $2.40, then the tip on 20% would be $4.80.

There is, however, one point that should be kept in mind: In many places in the United States, servers are paid a wage less than the minimum wage because tip income is estimated. That estimate is used to allow an employer to legally pay a wage rate less than the minimum wage because the tip income raises the total compensation above the minimum wage. In these cases, tip income is very important to these workers.

Sale Items

Everyone looks for and appreciates the reduction of the price of an item they wish to buy. However, all advertisements should not only be read with happiness but should also *raise a red flag*. It should be remembered that at times **sale items** are advertised in a media but are forgotten to be put into the computer. The sales representative simply scans the price on the item from the attached tag and pays little attention to what is put into the computer.

It is even possible for a salesperson to not be aware that an item is on sale. It is therefore necessary for the buyer to be aware of what the sales price should be and watch what price is registered when the salesperson scans it. This means that the buyer should be capable of determining the sales price of the item, or at least an approximation of the sales price, prior to the item being taken to the salesperson to ring up. The easiest way to determine this is to have a calculator. If not, an individual should be able to determine the approximate price mentally. The approximate price is mentioned because it is difficult to mentally determine the sales price and the tax.

Regardless of whether a calculator is used or the price is mentally determined, the following information may prove helpful.

Sales are offered at a 10%, 20%, or different percentage discounts. Sometimes on jewelry, a discount as high as 70% may be offered. An individual can use the same procedure as outlined for tipping to quickly determine the amount of a 10%, 15%, or 20% discount. The amount of a 25% discount can be determined by dividing the original cost of the item by 4. For example, if the original cost of an item is $20, dividing the $20 by 4 would result in a discount of $5.

When the reduction in the price of a sale item is determined, it is then necessary to reduce the original price of the item by this amount. However, if a calculator is used, or with reductions of 40% or more offered, it is generally easier to make the determination of the item a one-step procedure. The method to use to do this is to subtract the discount rate from 100% and then multiply. For example, if the discount rate is 40%, subtract the 40% from 100%, which will result in an amount of 60%. To find the sales price of the item, then multiply the original price by 6. For example, if the original price of an item was $30, move the decimal point one place to the left ($30 becomes $3) and multiply by 6. This means that the sales price would be 6 × 3, or $18.

One other factor on department store sales: Sometimes an advertisement will state a certain discount will be given and an additional amount will be given to those who buy it on the store's credit card (or for some other purpose). For example, the store may be offering a 40% discount on a set of dishes with an additional 10% discount for those using the store's credit card. This does not normally mean that the store is offering a 50% discount to those using the store's credit card. What it means is that a discount of 40% will be given and the sales price of the item determined. The 10% will then be given based on the determined sales price.

For example, an item is listed as having a sales price of 30% less than the regular price. The item normally sells for $140. With the 30% discount, the sales price will be $98. The 10% will then be given. This will be $9.80 and the item's final price will be $88.20.

Grocery Sale Items

Sales on groceries should raise a red flag even higher because it is more likely that a grocery sale item will not be put into the computer than a sale item at a department store. This is particularly true on grocery items that have been recently placed on sale.

It is normally not necessary to mentally determine the sales price of a grocery item. The sales price of the item is generally stated in the advertisement. The task of the buyer is to be aware of those items purchased that are on sale and be particularly alert to watch as those items are scanned. It is much easier for a clerk to check on an item as soon as the mistake has been called to his or her attention than it is to alert him or her to the mistake after the entire bill has been rung up. It is surprising how often a clerk is not aware of a sale item.

Reader's Practice Problems—9.1

1. Convert 57.6% to a decimal.
2. Convert .44 to a fraction and reduce the fraction to its lowest terms.
3. Convert 76% to a fraction and reduce the fraction to its lowest terms.
4. Convert .625 to a percent.
5. Convert $\frac{2}{5}$ to a percent.
6. Convert $7\frac{4}{7}$ to a percent.
7. What is 38% of 142?
8. The sales tax on a piece of luggage is 8.3%. A young lady bought a piece of luggage that cost $252.75. How much was the sales tax?
9. 52 is what percent of 170?
10. What percent of 625 is 30?
11. The discharge capability of a 2-inch tip is 995 gpm. The tip is discharging only 800 gpm. What percent of its full capability is being discharged from the tip?
12. The cost of a meal is $22.30. What would be the tip at a 10% rate, a 15% rate, and a 20% rate?
13. An $350 item is offered at a 22% discount. What would be the sales price of the item?
14. An $820 piece of jewelry is offered at a 60% discount. What would be the sales price of the item?
15. A TV is offered at a 25% discount with an additional 10% for the use of a department credit card. The sales tax on the TV is 7.5% What would be the total cost of the TV if the original price were $800?

The answers are found in the Reader's Practice Problems section of the Appendix.

REVIEW QUESTIONS

1. What is meant by a percent?
2. What numerical key is shared with the percent symbol?
3. How is a percent converted to a decimal?
4. How is a decimal converted to a fraction?
5. How is a percent converted to a fraction?
6. How is a decimal converted to a percent?
7. How is a fraction converted to a percent?
8. What procedure is used to determine the percent of a number?
9. Describe the 3 × 4 = 12 principle for solving alternate percent problems.
10. Define percentage.
11. What appears to be the most common percentages for tipping?
12. How is the amount of a 10% tip for a meal determined?
13. How is the amount of a 15% tip for a meal determined?
14. What is an easy method to use to determine the cost of a sale item that is offered at a 70% discount?
15. What warning should be remembered regarding sale items?

Test 9.1

1. Convert 73% to a decimal.
2. Convert 17.5% to a decimal.
3. Convert 24% to a fraction.
4. Convert 19% to a fraction.
5. Convert .31 to a percent.
6. Convert .735 to a percent.
7. Convert 3/5 to a percent.
8. Convert 5/11 to a percent.
9. What is 23% of 220?
10. A nozzle tip is capable of discharging 875 gpm. It is discharging only 82% of its capability. How much water is it discharging?
11. A nozzle tip that is capable of delivering 265 gpm is discharging only 185 gpm due to a reduction of the rpm's on the apparatus. What percent of its capability is the nozzle delivering?
12. A cylindrical tank that has a capacity of 6,000 gallons of water contains only 4,200 gallons. What percent of its capability is the tank holding?
13. 43 is 32% of what number?
14. 25 is 21% of what number?
15. If the total bill for a meal is $34.52 and the individual responsible for the bill decides to tip 15%, how much tip will be given?
16. The entire bill for a restaurant meal totaled $84.56. If the payer decides to leave a 20% tip, how much will the tip be?

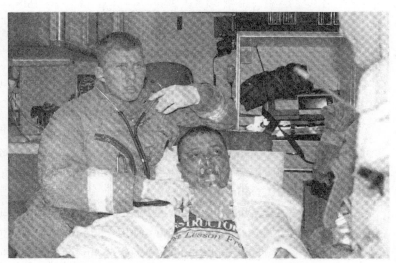

Figure 9.1

Courtesy of Rick McClure, LAFD

17. The regular price for a dress at a major department store is $82. The store is offering a 15% discount. How much will the sales price of the dress be?

18. A 60% discount is being offered on a diamond bracelet that normally sells for $1152. What would be the sales price for the bracelet?

19. A $230 item is offered for sale at a discount rate of 35%, with an additional 15% off for those buying between 10:00 a.m. and 12:00 noon. What would be the sales price for an individual who bought the item at 11:15 a.m.?

20. If an individual bought this same item on the day of the sale at 2:30 pm, what would be the sales price?

21. A report from the Rural Country Fire Department stated that 22% of its responses to medical emergencies were for difficult breathing problems similar to the one shown in Figure 9.1. If the department responded to 653 medical emergencies during the year, how many of them were for difficult breathing problems? (answer to the nearest whole number)

22. The Workhorse Fire Department responded to 789 medical emergency alarms last year. A total of 9% of them were to vehicle accident injuries similar to that shown in Figure 9.2. How many of this type of emergency did the department respond to last year? (answer to the nearest whole number)

Stop Solve the following three problems using the 3 × 4 = 12 concept.

23. What is 13% of 842?

24. 93 is 11% of what number?

25. 31 is what percent of 152?

26. At a single-family house fire, the firefighters find a victim with his chest and abdomen completely burned (18%) and both arms burned (9% each). What is the victim's total burn (bsa %)?

Figure 9.2

Courtesy of Rick McClure, LAFD

Chapter Summary

This chapter described the concepts of percent and percentages, and included problems designed for converting:

- a percent to a decimal
- a decimal to a fraction
- a percent to a fraction
- a decimal to a percent
- a fraction to a percent

The chapter also included a formula and sample problems on working with percentages, and explained how to not confuse the term percentile with percentage. A section on tipping and a section on the merits and hazards of sale items was also included, as well as problems for the reader to solve together with a series of review questions.

CHAPTER 10
WORKING WITH STATISTICS

KEY TERMS

central tendency
chart
data
datum
graph

horizontal axis
mean
median
mode
origin

skewed
standard curve of
   distribution
statistics
vertical axis

OBJECTIVES AND GOALS

Upon completion of this chapter, the reader should be able to:
- Define statistics, data, datum, central tendency, mode, median, mean, skewed, and graph.
- Determine the mode, median, and mean of a distribution.
- Describe a standard curve of distribution and explain the percentages of individuals who fall within one, two, and three standard deviations from the mean.
- Identify the various parts of a graph.
- Construct a chart or graph.
- Interpret information on a chart or graph.

WORDS TO REMEMBER

statistics—A branch of mathematics that is concerned with the collection, analysis, interpretation, and presentation of numerical data.

data—The plural of datum.

datum—A single piece of information.

central tendency—A means of measuring or describing the distribution of numerical values within a set of data.

mode—The value that most frequently appears in a distribution.

median—The middle value in a distribution when the values within the distribution have been arranged according to their magnitudes.

mean—The mathematic average value of the data in a set.

skewed—To misrepresent the true meaning or nature of something.

graph—A visual one- or two-dimensional figure that shows the relationship between two or more variables.

Introduction to Statistics

A wide definition of statistics is that it is the study of methods used to collect, present, analyze, and interpret quantitative data. A narrower definition is used in this book. In this book, statistics is defined as a branch of mathematics concerned with the collection, analysis, interpretation, and presentation of numerical data. Data is the plural of datum. Datum refers to a single piece of information.

In order to present the data collected when working with statistics as defined for this book, it is important to understand the three terms associated with central tendency. Central tendency is a means of measuring or describing the distribution of numerical values within a set of data by the use of a single value. The three most common measuring tools associated with this central tendency are the mean, median, and mode.

MEAN

The mean is also referred to as the arithmetic mean. It is the most commonly used unit of central tendency and is usually meant when reference is made to "the average" in common language. It is determined by adding all the values in a given set and dividing the result by the number of values in the data.

Problem: What is the mean for the distribution shown in Figure 10.1?

Solution: The total of the scores is 1,038. Because there are 12 scores in this distribution, it is necessary to divide the 1,038 by 12 to obtain the mean: $1038 \div 12 = 86.5$. Therefore, the mean for the distribution is 86.5.

Figure 10.1 Table of final test scores in EMT class

Student	Score
A	92
B	91
C	91
D	88
E	88
F	88
G	86
H	85
I	84
J	84
K	82
L	79

Figure 10.2 Table of yearly salaries for five workers

Worker #1	$42,000
Worker #2	$38,000
Worker #3	$37,000
Worker #4	$33,000
Worker #5	$29,000

Problem: What is the mean salary for the five workers shown in the distribution in Figure 10.2?

Solution: The total salaries of the five individuals is $179,000.
$179,000 divided by 5 = $35,800. Therefore, the average yearly salary of these five individuals is $35,800. This would be a fair assessment. However, a small sample of data, such as this, can be critically affected by an extremely high or low figure.

Problem: What is the mean average salary for the individuals shown in Figure 10.3?

Solution: The total of the yearly salaries shown in Figure 10.3 is $560,000. This amount divided by 7 (the total number in the sample) is:

$$\$560,000 \div 7 = \$80,000$$

This is not a true representation of the salaries listed. Such a result is said to be skewed.

MEDIAN

The median is the middle value in a distribution when the values within the distribution have been arranged according to their magnitudes. The distribution shown in Figure 10.1 has been so arranged. If there are an odd number of values in a distribution, the middle value is the median. If there are an even number of values within a distribution, the average of the two central items will be used as the median. The two central values in the distribution shown in Figure 10.1 are 88 and 86. Adding these two numbers and dividing by 2 produces an answer of 87. This is the median for the distribution shown in Figure 10.1.

Problem: Arrange the following data in a distribution according to their magnitude. Then determine the median.

23, 18, 21, 25, 14, 25, 44, 32, 33

Solution: Arranging the numbers according to their magnitude would produce the following distribution: 44, 33, 32, 25, 25, 23, 21, 18, 14. There are nine numbers in the distribution. The middle number is 25. This is the median for the distribution.

Figure 10.3 Table of yearly salaries for seven workers

Worker #1	$350,000
Worker #2	$42,000
Worker #3	$38,000
Worker #4	$37,000
Worker #5	$33,000
Worker #6	$31,000
Worker #7	$29,000

The median is used by a number of organizations to present a message to customers and the public. One example that is used often in both the written and visual medias is the sale price of homes in a given area. This data is watched closely when determining the trend of home values.

MODE

Mode is the value that most frequently appears in a distribution. It can be determined by scanning the numbers or scores in a distribution and selecting the one that appears most frequently.

Problem: What is the mode in the distribution shown in Figure 10.1?

Solution: Three students shown in the distribution received a score of 88. This is the score received by more students than any other score. Therefore, it is the mode of this distribution.

Problem: What is the mode in the following distribution?

$$25, 18, 19, 25, 31, 33, 84$$

Solution: Normally, the first step in answering a question such as this one is to arrange the numbers in a list from the highest to the lowest. The list would appear as:

84
33
31
25
25
19
18

There are two 25's. All other entries are single numbers. Therefore, 25 is the mode.

The Standard Curve of Distribution

The *standard curve of distribution* has no direct application to the gathering and distribution of fire service or EMS material. It is presented here primarily because of its close relationship to the study of statistics.

When a very large number of items appear in a distribution, the distribution tends to take on a normal shape. The shape that it takes is in the form of a bell and is generally referred to as a bell-shaped curve or the standard curve of distribution. In a true bell-shaped curve, the median, mode, and mean are at the central point of the curve and one side of the curve has the same shape as the opposite side. Figure 10.4a is a typical bell-shaped curve.

Figure 10.4a Standard curve of distribution

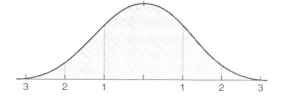

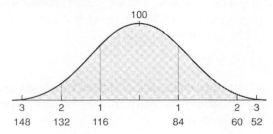

Figure 10.4b Distribution of IQ test scores

Each side of the bell-shape is divided into three parts. These parts are referred to as one standard deviation from the mean, two standard deviations from the mean, and three standard deviations from the mean. The standard deviation is calculated by whoever is responsible for presenting the data gathered for the study.

For a normal distribution of a set of data, the so-called empirical rule states that 68% of the elements of the data are within one standard deviation of the mean (34% on each side), and 95% of the elements of the data are within two standard deviations of the mean ($47\frac{1}{2}$% on each side), whereas 99.7% are within three standard deviations of the mean. The .3% that is not included within three standard deviations is generally referred to as tails.

Figure 10.4b is an indication of the distribution of IQ test scores. One standard distribution is 16. This means that on the test scores, 68% of those included in the datum for the distribution had an IQ between 84 and 116, 95% had a score between 68 and 132, and 99.7% had a score between 52 and 198. Very few people have an IQ above 198 or lower than 52.

Introduction to Graphs and Charts

When a fire service administrator submits a budget to the governing body, the budget normally contains the exact cost of all the items listed in the budget, including salaries. However, when the same information is submitted to the public through the written and visual medias, it is normally submitted as an approximate amount rather than the exact dollar figures. However, rather than submit information to the public regarding such items as the budget, number and types of alarms, fire prevention inspections, and so on, in written form, many administrators choose to use graphs or charts for illustrating these operations. Graphs and charts are considered a cousin to a photo, retaining the old saying that a picture is worth a thousand words. The objective of this chapter is to provide the reader with a basic understanding of the types and uses of these valuable tools.

For the purpose of this book, a ***graph*** or ***chart*** may be defined as a visual one- or two-dimensional figure that shows the relationship between two or more variables. A graph normally consists of a horizontal and vertical axis. The ***horizontal axis*** is referred to as the x-axis whereas the ***vertical axis*** is referred to as the y-axis. Each axis normally corresponds to one variable and they are labeled with different names. The location where the two axes intersect is called the ***origin***. It is also referred to as the point (0.0). The various parts of a graph are identified in Figure 10.5.

A point on the graph represents a relationship between the x-axis variable and the y-axis variable. The point is defined by a pair of numbers, one from the x-axis and one from the y-axis. For example, point A on Figure 10.6 would be identified as 20 on the x-axis and 30 on the y-axis.

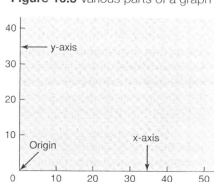

Figure 10.5 Various parts of a graph

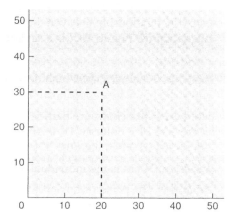

Figure 10.6 Point A on graph

Types of Graphs and Charts

Graphs and charts are very effective tools for presenting information quickly and easily. However, like courses on different subject matter and textbooks, graphs and charts should be used to fit the needs of a particular audience. Although pictographs, scatter plots, and hectographs are interesting to many different audiences, it is not believed that they have a particular use in the emergency services. Consequently, the introduction to graphs and charts in this chapter will be limited to bar graphs, line graphs, and pie charts.

Development of a Graph or Chart

When an administrator decides that the best method of presenting material to a particular audience is to use a graph or chart, certain principles should be kept in mind:

- Decide on the objective for using a particular type of graph or chart.
- Gather as much data as possible for presenting the material.
- Make sure that the information to be presented is accurate.
- Use as simple a design as possible, particularly if the material is complex.
- Present the material in such a manner that it will speak.
- Avoid any distortion or ambiguity.
- Integrate the style with text and tables.

Information for the Reader

A number of graphs and charts are illustrated in the remainder of this chapter. With the exception of the information in the chart shown in Figure 10.12 and the National Fire Protection Association information displayed in Figures 10.26, 10.27, and 10.28, all of the information in the graphs and charts is fictitious. It is included for the purpose of presenting types and uses of graphs and charts and not for the purpose of presenting useful information to the reader. Also, none of the organizations referred to in any of the graphs or charts have any relationship to any known organizations.

The charts shown in Figures 10.26, 10.27, and 10.28 were reprinted with permission from *NFPA Journal*®, September/October 2006, Copyright © 2006, National Fire Protection Association, Quincy, MA 02169.

BAR GRAPHS

A bar graph can be either a vertical bar graph or a horizontal bar graph. Both are used to compare the numerical values of several variables. The height or length of the bars indicates the relationship between the variables. Like pie charts, they are useful in comparing relationships. Figure 10.7 is a sample of a vertical bar graph.

Problem: What is the chart developer trying to illustrate in Figure 10.7?

Solution: The individual who developed the chart is trying to illustrate some of the primary reasons why occupants of buildings that have a fire do not leave the building.

Horizontal bar charts are particularly useful when the category names are too long to fit at the feet of a column. However, vertical bar graphs generally display data better and are preferred when possible. Figure 10.8 is a sample of a horizontal bar chart.

Problem: What is the individual who developed this bar chart trying to show?

Solution: The person who developed the chart is trying to illustrate the five most common reasons why individuals who have escaped from a building on fire reenter the building.

LINE GRAPHS

Because they are relatively easy to create and their visual characteristics reveal trends clearly, line graphs are more popular than all other types of graphs combined. Line graphs generally

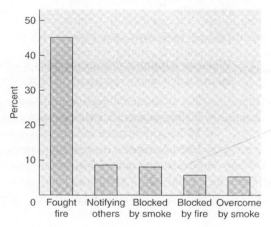

Figure 10.7

Reasons occupants did not leave fire building in XYZ County in 2006

Figure 10.8

The top five most frequent reasons why occupants reentered the fire building in XYZ County in 2006

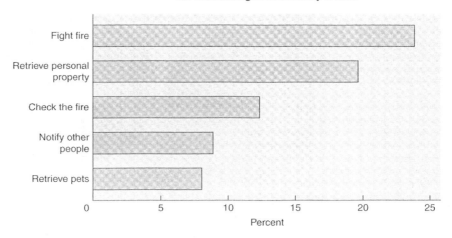

compare the relationship of one variable to another. However, a multiple-line graph can be used to effectively compare similar items over the same period of time. When a line graph compares only two variables, one of the variables is placed on the x-axis (horizontal axis) whereas the other is placed on the y-axis (vertical axis). If time is involved on the graph, it is normally placed on the x-axis whereas the quantity measured is placed on the y-axis. The chart shown in Figure 10.9 compares the number of responses to time periods and also illustrates that, in general, the number of emergency medical responses is on the increase.

Problem: Answer the following questions regarding information on the line graph shown in Figure 10.9.

1. Approximately how many medical emergencies did the XYZ County Fire Department respond to in 2005?
2. Approximately how many medical emergencies did the XYZ County Fire Department respond to in 2007?
3. From the information presented in the chart, does it appear that the number of medical alarms will increase, decrease, or stay the same in 2009?

Figure 10.9

Medical emergency responses by XYZ County Fire Department 2002–2006

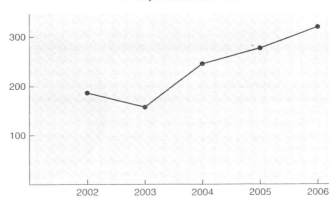

Figure 10.10

Total number of ambulance responses in one week

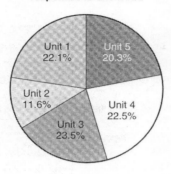

Answers:

1. About 155
2. About 270
3. Increase

PIE CHARTS

Pie charts are generally used when it is desired to show the percentage parts of a whole. An example of this is illustrated in Figure 10.10. Figure 10.10 indicates the percentage of alarms that five different ambulance units made in a one-week period as compared with the total number of ambulance response alarms made in a one-week period.

At times a particular segment is separated from the main drawing in order to emphasize an important piece of information. An example of this situation is shown in Figure 10.11.

A pie chart also has another use. For lack of a better term, it will be referred to as a constant changing pie chart. It can be found on any computer under the term "properties" for the hard drive (C drive). Information on the pie chart will change whenever anything new is added to or removed from the hard drive. Check it out. It is worth reviewing occasionally, particularly if a hard drive is getting close to being full. If this happens, it might be wise to consider removing some of the material from the hard drive

Figure 10.11

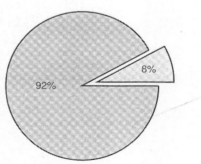

Figure 10.12

■ Used space 13, 273, 276, 416 bytes 12.3 GB
☐ Free space 141, 419, 560, 968 bytes 131.7 GB
Capacity 154, 692, 837, 376 bytes 144 GB

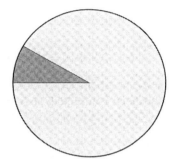

because it has a direct influence on how fast a computer works. A sample of the information for a relatively new computer is illustrated in Figure 10.12.

REVIEW QUESTIONS

1. What is the definition of statistics?
2. What is the definition of data?
3. What is the definition of datum?
4. What is the definition of central tendency?
5. What is the definition of mode?
6. What is the definition of median?
7. What is the definition of skewed?
8. What is the definition of mean?
9. How is the median determined when there is an even number of values in a distribution?
10. Which of the central tendency terms is normally used when "the average" is referred to in normal language?
11. If an extremely high or low value is used in a distribution, what normally happens to the mean?
12. According to the empirical rule, what percent of the elements of a distribution that fall within a standard curve of distribution are within one standard deviation of the mean?
13. What is the meaning of the phrase that graphs and charts are looked on as a cousin to a photo?
14. For the purpose of this book, what is the definition of a graph or chart?
15. Which of the two axes on a graph is referred to as the x-axis?
16. What points should an EMS administrator keep in mind when developing a graph or chart?

17. What type of bar chart generally displays data better?
18. What type of chart is the most popular?
19. When are pie charts generally used?
20. Where might a constant changing pie chart be found?

Test 10.1

78, 52, 96, 4, 17, 82, 97, 45, 23, 78, 33, 52, 77, 95, 52, 33, 95, 52, 90

Place the numbers shown above in a weighted sequence from highest to lowest. When this has been completed, answer Problems 1, 2, and 3.

1. What is the median of this distribution?
2. What is the mode?
3. What is the mean?
4. What is the median of the miles traveled in the Figure 10.13 distribution?
5. What is the mode of the miles traveled in the Figure 10.13 distribution?
6. What is the mean of the miles traveled in the Figure 10.13 distribution?

Figure 10.13 Emergency responses for Engine 4 in June

Response Number	Date	Miles Traveled
1	2	2.3
2	3	1.8
3	5	0.9
4	7	2.3
5	8	1.9
6	9	3.4
7	10	2.1
8	11	1.9
9	12	3.1
10	13	2.9
11	17	1.9
12	19	0.8
13	20	2.1
14	21	3.7
15	22	1.1
16	24	1.3
17	25	2.2
18	27	4.5
19	27	1.9
20	28	2.2
21	29	1.7
22	30	1.9

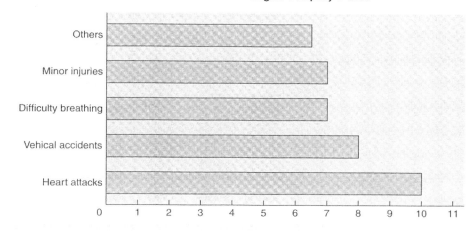

Figure 10.14

Medical emergency incidents responded to by volunteer Engine Company 1–2006

7. According to the information illustrated in Figure 10.14, what type of medical emergency did Engine Company 1 respond to most often in 2006?

8. According to the information illustrated in Figure 10.14, what two types of medical emergencies did Engine Company 1 respond to about the same number of times in 2006?

9. According to the information illustrated in Figure 10.15, what type of action is a male occupant of a building most likely to take first in the event of a fire?

10. According to the information illustrated in Figure 10.15, approximately what percent of the time is a male occupant of a building likely to call the fire department as his first type of action?

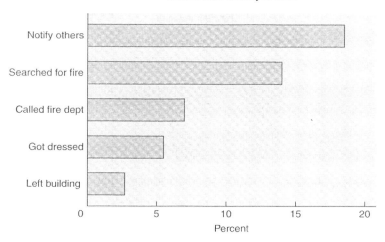

Figure 10.15

First actions of male occupants to fires in XYZ County in 2006

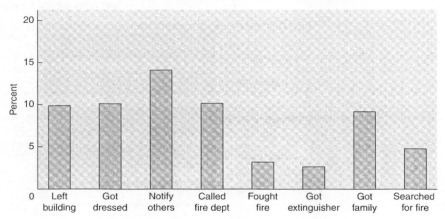

Figure 10.16

First actions of female occupants to fires in XYZ County in 2006

11. According to the information illustrated in Figure 10.16, what type of action is a female occupant of a building most like to take first in the event of a fire?
12. Using the information in Figures 10.15 and 10.16, is a male or female most likely to call the fire department first?
13. Using the information in Figures 10.15 and 10.16, is a male or female most likely to get dressed first?
14. If the bar graph in Figure 10.17 represents 1,348 fires in property-carrying vehicles, approximately how many of the fires were caused by electrical equipment?
15. If the line graph in Figure 10.18 represents that there were 895 structure fires in XYZ County in 2005, approximately how many of these fires were either incendiary or suspicious?
16. If the bar graph in Figure 10.19 represents 5,482 fires, approximately how many of these fires involved upholstery furniture?
17. According to the graph illustrated in Figure 10.20, approximately how much was the fire loss in XYZ County in 2006?

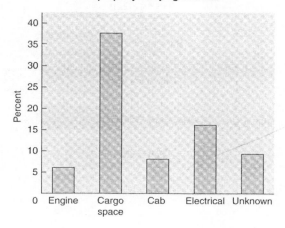

Figure 10.17

Primary location of origin of fire in property-carrying vehicles

WORKING WITH STATISTICS ■ 237

Figure 10.18

Percent of structure fires in XYZ County that were incendiary or suspicious 2002–2006

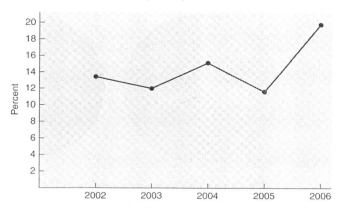

Figure 10.19

Percent of fires involving textiles as the primary fuel that resulted in the fatality

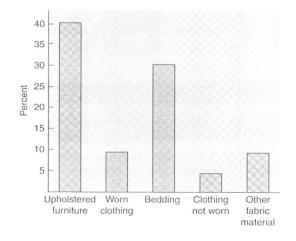

Figure 10.20

Trend for fire loss in XYZ County 2002–2006

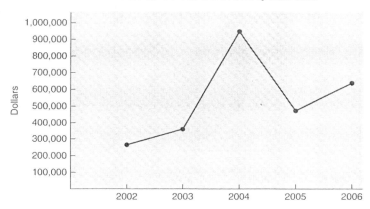

Figure 10.21

Types of alarms in a one year period for Paramedic Unit 1

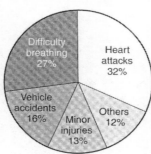

18. According to the information in Figure 10.21, which two types of incidents accounted for 29% of Paramedic Unit 31's responses in a one-year period?
19. According to the information in Figure 10.22, what percent of the responses in a six-month period for Volunteer Fire Department XYZ was for fires?
20. According to the information in Figure 10.23, approximately what percent of the fires in passenger vehicles were located in the trunk, brakes, and fuel tank combined?
21. If the information in Figure 10.24 was for 652 civilian fire deaths in XYZ State in 2006, how many deaths were caused by smoking and matches?
22. If the information in Figure 10.25 represents 2,388 fires, how many of these were wildfires?
23. Which of the types of fires indicated on the chart in Figure 10.26 had the most steady trend between 1977 and 2000?
24. According to the information in Figure 10.26, what type of fire was responsible for about one-half of the total number of fires in 1977?
25. According to the information in Figure 10.26, what year between 1985 and 1995 had the most total fires?

Figure 10.22

Types of alarms in a six-month period for Volunteer Fire Department XYZ

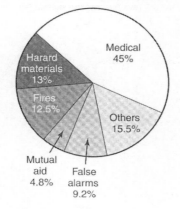

Figure 10.23

Area of fire origin in passenger car fires

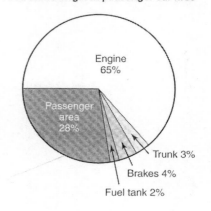

Figure 10.24 Pie chart: causes of civilian deaths

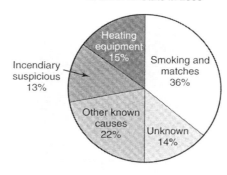

Percentage causes of civilian fire deaths in the XYZ State in 2006

Figure 10.25

Fires by location 2006

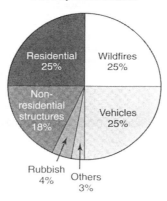

Figure 10.26

Estimates of fires by type in the United States (1977–2005)

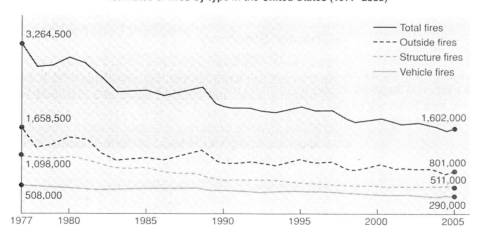

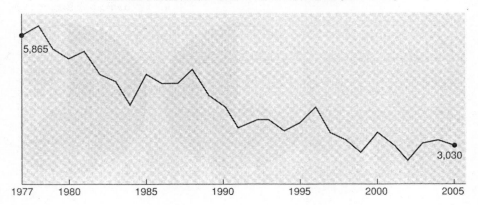

Figure 10.27

Civilian fire deaths in the home in the United States (1977–2005)

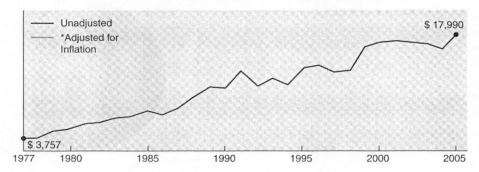

Figure 10.28

Average property loss per structure* fire in the United States (1977–2005)

26. From the information in Figure 10.27, in what year between 1991 and 2005 were there more civilian fire deaths in the home?

27. Figure 10.28 compares the average property loss per structure fire in the United States between 1977 and 2000 by the loss adjusted for inflation and the unadjusted loss. Which of the two had the most steady trend?

Chapter Summary

This chapter introduced the principles of statistics and the use of graphs and charts, as well as defined words commonly associated with statistics. It also provided examples of how the mean, median, and mode of a distribution are identified and determined. The use of graphs and charts were explained, together with the important points to be considered when preparing a graph or chart. Numerous problems were presented to evaluate the reader's knowledge of the information in the chapter, and review questions were provided.

CHAPTER 11
FORMULAS AND FORMULA MANIPULATIONS

KEY TERMS
equation   formula

OBJECTIVES AND GOALS
Upon completing this chapter, the reader should be able to:
- Define the words formula and equation.
- Manipulate fire department hydraulic formulas.
- Solve fire department hydraulic formulas.

WORDS TO REMEMBER

formula—A pathway designed to obtain the answer to a mathematical problem. Formulas are used extensively in the fire service for solving data related to book hydraulic and field problems.

equation—A mathematical statement wherein two expressions separated by an equal sign are of the same value.

Solving Formula Problems

Formulas used in the emergency services generally have one or more letters to the left side of an equal sign that represent the information to be determined (solved). If the information that is desired to be determined is on the right side of the equal sign, then it is necessary to manipulate the formula to change it to the left side of the equal sign. As an example, in its simplest form, the formula used for determining the required pump discharge pressure for a pumper supplying water to a hose line is:

$$RPDP = NP + FL$$

where   RPDP = required pump discharge pressure
         NP = nozzle pressure
         FL = friction loss

If the nozzle pressure and the friction loss are known, the required pump discharge pressure can be determined by adding the two. However, the problem is different if the RPDP and the nozzle pressure are known, and it is necessary to determine the friction loss. The problem becomes easy once the formula is manipulated (changed) to provide a new formula for determining the friction loss when the RPDP and the nozzle pressure are known. The new formula would be:

$$FL = RPDP - NP$$

The primary objective of this chapter is to explain the procedure required for manipulating a formula to obtain a new formula in order to determine the information desired. The common formulas provided for solving hydraulic problems in the fire service will be used for this purpose. A series of problems provided at the end of the chapter then tests the ability of the reader to accept the challenge.

Manipulating Formulas

Several factors should be kept in mind when manipulating a formula and preparing to develop a new formula to use in determining the desired answer:

1. Everything on one side of an equal sign is equal to everything on the other side of the equal sign.
2. Anything can be done to one side of the equal sign as long as the same thing is done to the other. This is an extremely important point to remember. However, whatever is done *must be done to the entire side, not a portion of the side*. For example, a number or letter can be added to one side of the equal sign as long as the same number or letter is added to the other side.

Example:

$$8 - 3 = 5$$

Now add 3 to both sides.

$$8 - 3 + 3 = 5 + 3$$

This cancels out the -3 on the left side.

Result: $8 = 8$

Problems and formulas that directly involve fire department hydraulics will be used to explain the procedure for manipulating formulas. Each problem will be illustrated by a drawing. The procedure for manipulating the formula basically involves moving four different types of items from the left side of the equal side to the right side. The four basic items that might be involved are:

1. Positive and negative units; for example: $a + b - c$
2. Multiplication units; for example: $(a)(b)(c)$
3. Exponent units; for example: A^2
4. Square root units; for example: $\sqrt{A}$

Working with Positive and Negative Units

Problem: What is the friction loss in the hose lines in the layout shown in Figure 11.1?

Solution: The formula for determining the required pump discharge pressure (RPDP) when lines are laid at ground level into a portable monitor is:

Figure 11.1

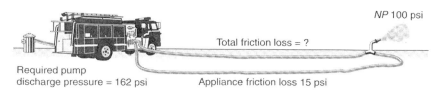

$$RPDP = NP + FL + AFL$$

where   RPDP = required pump discharge pressure
NP = nozzle pressure
FL = total friction loss in the hose lines
AFL = appliance friction loss

The first step in the manipulation of any formula is to move the unknown factor to the left side of the equal sign. This is done by switching the information in the given formula. The formula RPDP = NP + FL + AFL would then appear as:

$$NP + FL + AFL = RPDP$$

Then it is necessary to move the items on the left side of the equal sign that are with the item to be solved to the right side of the equal sign.

The objective of this sample problem is to set up a formula for determining the FL. Therefore, the NP and the AFL on the left side need to be moved to the right side. This involves an important principle to remember: *When all letters or digits on the left side of the equal sign are positive or negative, any of them can be moved to the right side by simply changing its sign.* This means that the NP and the AFL can be moved from the left side of the equal sign to the right side by simply changing their sign. The result would appear as:

$$FL = RPDP - NP - AFL$$

In effect, this was achieved by adding a $-NP$ and a $-AFL$ to both sides of the equal sign. The operation would appear as:

$$NP + FL + AFL - NP - AFL = RPDP - NP - AFL$$

then cancel

$$NP + FL + AFL - NP - AFL = RPDP - NP - AFL$$

and $FL = RPDP - NP - AFL$

The RPDP is shown in Figure 11.1 to be 162 psi, the nozzle pressure is shown to be 100 psi, and the appliance friction loss is 15 psi. This information can be placed into the new formula, and the formula can be used to solve the problem. The process would appear as:

$$FL = 162 - 100 - 15$$
$$= 47 \text{ psi}$$

Problem: How much is the back pressure in the layout illustrated in Figure 11.2 (see page 244)?

Solution: The formula for determining the RPDP when lines are laid at ground level into a ladder pipe is:

$$RPDP = NP + FL + BP + AFL$$

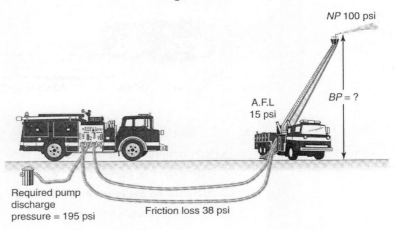

Figure 11.2

where RPDP = required pump discharge pressure
NP = nozzle pressure
FL = friction loss
BP = back pressure
AFL = appliance friction loss

To manipulate the formula, first change the items to the opposite side. The result is:

$$NP + FL + BP + AFL = RPDP$$

The objective is to set up a formula to solve for the BP. To do so, it is necessary to move the NP, the FL, and the AFL to the right side of the equation. The result would appear as:

$$BP = RPDP - NP - FL - AFL$$

To solve the problem, replace the letters with the known numbers from Figure 11.2.

$$BP = 195 - 100 - 38 - 15$$
$$= 42 \text{ psi}$$

Working with Multiplication Units

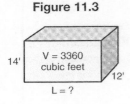

Figure 11.3

Problem: What is the length of the rectangular container shown in Figure 11.3?

Solution: The formula for determining the volume of a rectangular container is:

$$V = l \, w \, h$$

where V = volume
l = length
w = width
h = height

First, switch the items to the opposite side of the equal sign so that the item to be solved is on the left side. The result is:

$$l \, w \, h = V$$

The objective is to set up a formula to solve for l. This means that the "w" and "h" need to be moved to the right-hand side of the equal sign. When all items on the left side of the equal sign are being multiplied, those that need to be moved can be moved by dividing both sides of the equal sign by those items. The operation would appear as:

$$\frac{(l)(w)(h)}{(w)(h)} = \frac{V}{(w)(h)}$$

The (w) and (h) on the left side of the equal sign can now be canceled out.

$$\text{cancel} \downarrow\text{---}\downarrow$$
$$\frac{(l)(w)(h)}{(w)(h)} = \frac{(V)}{(w)(h)}$$
$$\text{cancel} \uparrow\text{---}\uparrow$$

This results in the formula: $l = \dfrac{V}{(w)(h)}$

where V = 3,360 cubic feet
       w = 12 feet
       h = 14 feet

Applied to the developed formula, it appears as:

$$l = \frac{3{,}360}{(12)(14)}$$
$$= \frac{3{,}360}{168}$$
$$= 20 \text{ feet}$$

Problem: What is the head (H) in Figure 11.4?

Solution: The formula for determining the pressure when the head is known is:

$$P = .434\,H$$

where P = pressure in psi
       H = head

Figure 11.4

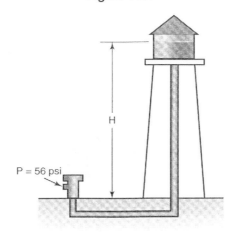

Step 1. Switch items to the opposite side of the equal sign.

Result: .434 H = P

Step 2. Divide both sides by .434.

$$\frac{(.434)(H)}{(.434)} = \frac{P}{.434}$$

Step 3. Cancel the .434 on the left side of the equal sign.

$$\frac{(.434)(H)}{(.434)} = \frac{P}{.434}$$

Result: $H = \dfrac{P}{.434}$

Step 4. Solve the problem: $H = \dfrac{56}{.434}$

$$H = 129.03 \text{ feet}$$

Working with Exponent Units

Problem: Remove the exponent (2) from S^2.

Solution: A square can be eliminated by taking the square root.

Example: $\sqrt{S^2} = S$

Problem: What is the flow of water in the line shown in Figure 11.5?

Solution: To determine the amount of water flowing through the line, it is necessary to manipulate the friction loss formula.

The formula for determining the friction loss in a layout is:

$$FL = CQ^2L$$

where FL = friction loss in the layout

C = 2 (for $2\frac{1}{2}$-inch hose)

Q = flow in hundreds of gallons (flow/100)

L = length in hundreds of feet (length/100)

Step 1. Switch the items to the opposite side of the equal sign.

$$CQ^2L = FL$$

Figure 11.5

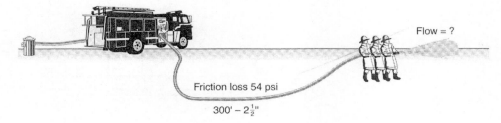

Step 2. Divide both sides by CL.

$$\frac{CQ^2L}{CL} = \frac{FL}{CL}$$

Step 3. Cancel the C and L on the left side of the equal sign.

cancel--------↓---↓
$$\frac{CQ^2L}{CL} = \frac{FL}{CL}$$
cancel--------↑-↑

Result: $Q^2 = \dfrac{FL}{CL}$

Step 4. It is now necessary to remove the square from the Q. This can be done by taking the square root of both sides of the equal sign.

$$\sqrt{Q^2} = \sqrt{FL/CL}$$

Step 5. The Q^2 becomes Q. $\quad Q = \sqrt{FL/CL}$

Step 6. Enter the known figures.
$$\begin{aligned}Q &= \sqrt{54/(2)(3)} \\ &= \sqrt{54/6} \\ &= \sqrt{9} \\ &= 3\end{aligned}$$

Step 7. Change the 3 to gpm by multiplying by 100.

$$3 \times 100 = 300 \text{ gpm}$$

Working with Square Root Units

Problem: Remove the square root from the C.

Solution: A square root can be eliminated by squaring the unit.

Example: $(\sqrt{C})^2 = C$

Problem: What is the head in the situation shown in Figure 11.6?

Solution: The formula for velocity flow is:

$$V = 8\sqrt{H}$$

Figure 11.6

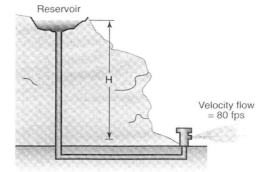

where V = velocity flow in fps (feet per second)
H = head

Step 1. Switch the units to the opposite side of the equal sign.

$$8\sqrt{H} = V$$

Step 2. Divide both sides by 8.

$$\frac{8\sqrt{H}}{8} = \frac{V}{8}$$

Step 3. Cancel the 8's on the left side of the equal sign.

cancel --------↓
$$\frac{8\sqrt{H}}{8} = \frac{V}{8}$$
cancel ----------↑

Result: $\sqrt{H} = \dfrac{V}{8}$

Step 4. Square both sides.

$$(\sqrt{H})^2 = (V/8)^2$$

Result: $H = (V/8)^2$

Step 5. Enter the V value.

$$H = (80/8)^2$$

Step 6. Solve the problem.

$$H = (80/8)^2$$
$$= 10^2$$
$$= 100 \text{ feet}$$

Double Trouble Situations

Some fire department hydraulic formulas require the application of more than just one of the formula manipulation principles that were previously examined:

- When all letters or digits on the left side of the equal sign are positive or negative, any of them can be moved to the right side by simply changing its sign.
- When all items on the left side of the equal sign are being multiplied, those that need to be moved can be moved by dividing both sides of the equal sign by those items.
- A square can be eliminated by taking the square root of the unit.
- A square root can be eliminated by squaring the unit.

For lack of a better term, formulas that require the application of more than one of the above listed principles can be defined as double trouble formulas. Manipulating these formulas is sometimes complex. Following is a sample of one of these formulas that illustrates how the manipulation of such a formula is solved.

Problem: The formula for determining Centigrade is:

$$C = \frac{(5)(F - 32)}{9}$$

Convert this formula to a formula for determining Fahrenheit.

Solution:

Step 1. Switch the units to the opposite side of the equal sign.

$$\frac{(5)(F - 32)}{9} = C$$

Step 2. Multiply both sides by 9.

$$\frac{(9)(5)(F - 32)}{9} = 9C$$

Step 3. Cancel the 9's on the left side of the equal sign.

$$\text{cancel} \downarrow$$
$$\frac{(9)(5)(F - 32)}{9} = 9C$$
$$\text{cancel} \uparrow$$

Result: $(5)(F - 32) = 9C$

Step 4. Divide both sides by 5.

$$\frac{(5)(F - 32)}{5} = \frac{9C}{5}$$

Step 5. Cancel the 5's on the left side of the equal sign.

$$\text{cancel} \downarrow$$
$$\frac{(5)(F - 32)}{5} = \frac{9C}{5}$$
$$\text{cancel} \uparrow$$

Result: $F - 32 = \dfrac{9C}{5}$

Step 6. Move the 32 from the left side to the right side and change its sign. The result is the required formula:

$$F = \frac{9C}{5} + 32$$

Problem: What is the nozzle pressure on the hose line shown in Figure 11.7?

Solution: The formula for determining the discharge from a nozzle tip is:

$$\text{Dis} = 29.7 \, D^2 \sqrt{P}$$

where Dis = discharge from the nozzle tip
 D = diameter of the nozzle tip
 P = nozzle pressure

Figure 11.7

Discharge 328 gpm
Tip $1\frac{1}{4}"$
NP = ?

Note that this formula has both a letter being squared and the square root being determined on another letter. Both the discharge and the size of the tip being used are shown in Figure 11.7. The discharge is 328 gpm and the tip size is $1\frac{1}{4}$ inch, which should be used in the formula as 1.25.

Step 1. Switch the units to the opposite sides.

$$29.7\ D^2\sqrt{P} = Dis$$

Step 2. Divide both sides by 29.7 D^2.

$$\frac{29.7\ D^2\sqrt{P}}{29.7\ D^2} = \frac{Dis}{29.7\ D^2}$$

Step 3. Cancel the 29.7 D^2 on the left side.

cancel --------↓---↓
$$\frac{29.7\ D^2\sqrt{P}}{29.7\ D^2} = \frac{Dis}{29.7\ D^2}$$
cancel--------↑--↑

Result: $\sqrt{P} = \dfrac{Dis}{29.7\ D^2}$

Step 4. Square both sides.

$$(\sqrt{P})^2 = (Dis/29.7\ D^2)^2$$

Result: $P = (Dis/29.7\ D^2)^2$

This is the formula sought.

Step 5. Inject the known figures into the formula including squaring the D.

$$P = [328/(29.7)\ (1.25)\ (1.25)]^2$$

Step 6. Solve the problem.

$$= (328/46.41)^2$$
$$= (7.07)^2$$
$$= 49.98\ \text{or}\ 50\ \text{psi}$$

Before subjecting you to the review questions, it is probably best to expose you to a few word problems to see if you understood the manipulating process just explained. Work each of the following problems. If you have any trouble, go back and review the section that explains exactly how you should have worked the problem. Some readers may have a little trouble with this chapter because it led you into thinking of algebra rather than ordinary math.

Remember, you will be working on formula manipulation problems. The first step should always be to manipulate the formula to provide for a new formula that will solve the problems.

Problem 1. Your crew is working on a line that is used to extinguish a pile of trash that is located above the pump. The nozzle pressure on the line being used is 45 psi and the friction loss in the line is 25 psi. The pump pressure supplying the line is 190 psi. How far above the pump is the pile of trash that is being extinguished?

SOLUTION HELP—This is a two-step problem. First, use the formula RPDP = FL + NP + BP to determine the BP. Then use the formula BP = .434 H to determine how far above the pump the fire is located.

Problem 2. The volume of a rectangular container is 576 square feet. The container measures 12 feet by 8 feet by an unknown height. What is the height?

The formula for determining the volume of a rectangular container is:

$$V = l \, w \, h$$

where   l = length
w = width
h = height

Problem 3. The reach of a horizontal stream from a handheld nozzle is 72.11 feet. The horizontal factor for the nozzle is 104. What is the pressure on the nozzle?

The formula for determining the reach of the stream is:

$$S = \sqrt{(HF)(P)}$$

where   HF = horizontal factor
P = nozzle pressure

Problem 4. The velocity flow of the water issuing from a hole in the side of a tank that is located some distance from the surface of the water in the tank is 96 feet per second. How far below the surface of the water is the hole located?

The formula for determining the velocity flow is:

$$V = 8\sqrt{H}$$

where   H = head in feet

Problem 5. A $1\frac{1}{4}$-inch tip is being used on the end of a line. The nozzle reaction from the water leaving the line is 147.19 pounds. What is the nozzle pressure on the tip?

The formula for determining the nozzle reaction is:

$$NR = 1.57 D^2 P$$

where   D = nozzle diameter
P = nozzle pressure

Problem 6. Engine 5's crew are working at the end of a $2\frac{1}{2}$-inch line. The discharge from the tip on their line is 209.98 gpm. The discharge pressure is 50 psi. What is the size of the tip?

The formula for determining the discharge from a tip is:

$$\text{Discharge} = 29.7 D^2 \sqrt{P}$$

where   D = diameter of the tip
P = discharge pressure ($\sqrt{50} = 7.07$)

Problem 7. The layout is to a portable monitor. The line is laid up hill that is providing a back pressure of 46 psi. The appliance friction loss on the monitor is 15 psi and the friction loss in the line is 57 psi. The engineer is pumping at 198 psi in order to supply the layout. What is the nozzle pressure on the monitor?

The formula for determining the pressure that the engineer has to supply for the layout is:

$$RPDP = NP + FL + BP + AFL$$

where   RPDP = required pump discharge pressure
NP = nozzle pressure
FL = friction loss

BP = back pressure

AFL = appliance friction loss

Problem 8. The formula for determining the amount of water discharge from a sprinkler head is:

$$\text{Discharge} = (K)(\sqrt{P})$$

where   K = the coefficient of the sprinkler head

P = Discharge pressure

The sprinkler system in a local hardware store discharges water at the rate of 22 pounds (psi). The heads have a coefficient of 8.2. What would be the discharge pressure on the heads in the system?

Problem 9. The formula for determining the discharge from a nozzle is:

$$\text{Discharge} = 29.7D^2\sqrt{P}$$

where   D = diameter of the nozzle tip

P = nozzle pressure

The nozzle being used at the end of a $1\frac{1}{2}$-inch line is operating at 80 psi nozzle pressure and discharging 66 gpm. What size tip is being used on the line?

Problem 10. The formula for determining the FL on a single line is:

$$FL = CQ^2L$$

where   FL = friction loss in the layout

C = friction loss coefficient

Q = flow in hundreds of gpm (flow/100)

L = hose length in hundreds of feet (length/100)

Engine 3's crew is operating a nozzle on the end of a 2-inch line. The length of the line from the pumper to the fire is 250 feet. The friction loss coefficient on the 2-inch line is 8. The total friction loss in the hose layout is 31.25 psi. What is the amount of water flowing through the line?

REVIEW QUESTIONS

1. What is the definition of a formula?
2. What is the definition of an equation?
3. What are two points to remember regarding the manipulation of a formula?
4. What is an important point to remember when changing positive or negative units from one side of the equal sign to the other?
5. When all units on the left side of the equal sign are being multiplied, how are units that are to be moved to the other side processed?
6. When a unit is being squared, how can the squared sign be removed to result in the unit being by itself?
7. When the square root of a unit is indicated, how can the square root sign be removed?
8. What is generally the first step taken when manipulating a formula?

Reader's Practice Problems—11.1

Note: The answers, with a complete demonstration of how each answer was obtained, are provided in the Reader's Practice Problems section of the Appendix.

1. The formula for determining the head when the pressure is known is H = 2.304 P. Manipulate this formula to develop a formula for determining the pressure when the head is known.

2. The formula for determining the nozzle reaction on a nozzle is NR = $1.57D^2P$. Manipulate this formula to develop a formula for determining the diameter of the nozzle tip when the nozzle reaction and the nozzle pressure are known.

3. The discharge formula is:

$$\text{Discharge} = 29.7\, D^2 \sqrt{P}$$

   Manipulate this formula to develop a formula for determining the pressure when the discharge and the diameter are known.

4. The formula for determining the required pump discharge pressure when lines are laid downhill into a deck gun is RPDP = NP + FL + AFL − FP. Manipulate this formula to determine the FL when the RPDP, NP, AFL and FP are known.

5. The formula for determining the velocity flow when the head is known is V = 8$\sqrt{H}$. Manipulate this formula to develop a formula to determine the head when the velocity flow is known.

6. One of the formulas for determining the horizontal reach of a fire stream is S = P/2 + 15. Manipulate this formula to develop a formula to determine the pressure when the horizontal reach is known.

7. The formula for determining the forward pressure (FP) when the head is known is:

$$FP = .434\, H$$

   What is the head in Figure 11.8?

Figure 11.8

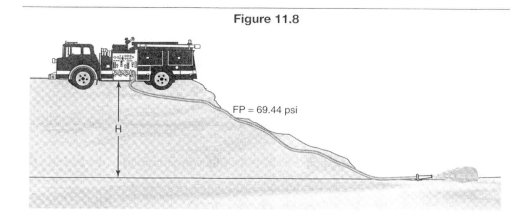

FP = 69.44 psi

Figure 11.9

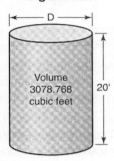

Figure 11.10

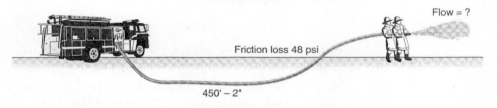

Figure 11.11

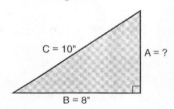

8. The formula for determining the volume of a cylindrical container is:

$$V = .7854 D^2 H$$

   What is the diameter of the container shown in Figure 11.9?

9. The formula for determining the friction loss in a hose line is:

$$FL = CQ^2L$$

   where   C for a 2-inch hose is 8
                Q = flow/100
                L = length of line/100

   What is the flow (Q) in Figure 11.10?

10. The formula for finding side C in a right triangle is:

$$C\sqrt{A^2 + B^2}$$

   What is A in Figure 11.11?

TEST 11

Manipulate each of the following formulas to determine a new formula for what is asked.

1. The formula: volume = $.7854 D^2 H$

   Develop a formula for D.

FORMULAS AND FORMULA MANIPULATIONS ■ 255

SOLUTION HELP

Step 1: Change the sides of the formula. $.7854D^2H = \text{volume}$

Step 2: Divide both sides by .7854 H $\dfrac{.7854D^2\ H}{.7854\ \ H} = \dfrac{\text{volume}}{.7854\ H}$

cancel---↓--------↓

Step 3: Cancel out .7854 H on the left side. $\dfrac{.7854D^2\ H}{} = \dfrac{\text{volume}}{.7854\ H}$

cancel----↑-------↑

$= D^2 = \text{volume}/.7854$

Step 4: Take the square root of both sides.

$$\sqrt{D^2} = \sqrt{\text{volume}/.7854\ H}$$
$$= D = \sqrt{\text{volume}/.7854\ H}$$

Step 5: The formula is developed. $D = \sqrt{\text{volume}/.7854\ H}$

2. The formula: weight capacity $= 62.5\ V$
Develop a formula for V.

SOLUTION HELP

Step 1: Change the sides of the formula. $62.5\ V = \text{weight capacity}$

Step 2: Divide both sides by 62.5. $\dfrac{62.5\ V}{62.5} = \dfrac{\text{weight capacity}}{62.5}$

cancel-↓

Step 3: Cancel 62.5 on the left side. $\dfrac{62.5\ V}{62.5} = \dfrac{\text{weight capacity}}{62.5}$

cancel-↑

Step 4: The formula is developed. $V = \dfrac{\text{weight capacity}}{62.5}$

3. The formula: $S = \sqrt{(HF)(P)}$
Develop a formula for P.

SOLUTION HELP

Step 1: Change the sides of the formula. $\sqrt{(HF)(P)} = S$

Step 2: Square both sides. $[\sqrt{(HF)(P)}]^2 = S^2$

$= (HF)(P) = S^2$

Step 3: Divide both sides by HF. $\dfrac{(HF)(P)}{HF} = \dfrac{S^2}{HP}$

cancel-↓

Step 4: Cancel HF on the left side. $\dfrac{(HF)(P)}{HF} = \dfrac{S^2}{HP}$

cancel-↑

Step 5: The formula is developed. $P = \dfrac{S^2}{HP}$

4. The formula: $V = 12.14\sqrt{P}$
   Develop a formula for P.
   SOLUTION HELP
   Step 1: Change the sides of the formula. $\quad 12.14\sqrt{P} = V$

   Step 2: Divide both sides by 12.14. $\quad \dfrac{12.14\sqrt{P}}{12.14} = \dfrac{V}{12.14}$

   cancel----↓
   Step 3: Cancel 12.14 on the left side. $\quad \dfrac{12.14\sqrt{P}}{12.14} = \dfrac{V}{12.14}$
   cancel---↑

   $\sqrt{P} = V/12.14$

   Step 4: Square both sides. $\quad (\sqrt{P})^2 = (V/12.14)^2$

   $P = (V/12.14)^2$

   Step 5: The formula is developed. $\quad P = (V/12.14)^2$

5. The formula: $VD^2 = vd^2$
   Develop a formula for v.
   SOLUTION HELP
   Step 1: Change the sides of the formula. $\quad vd^2 = VD^2$

   Step 2: Divide both sides by d^2. $\quad \dfrac{vd^2}{d^2} = \dfrac{VD^2}{d^2}$

   cancel-↓
   Step 3: Cancel d^2 on the left side. $\quad \dfrac{vd^2}{d^2} = \dfrac{VD^2}{d^2}$
   cancel-↑

   Step 4: The formula is developed. $\quad v = \dfrac{VD^2}{d^2}$

6. The formula: $Q = K\sqrt{P}$
   Develop a formula for P.
   SOLUTION HELP
   Step 1: Change the sides of the formula. $\quad K\sqrt{P} = Q$

   Step 2: Divide both sides by K. $\quad \dfrac{K\sqrt{P}}{K} = \dfrac{Q}{K}$

   cancel-↓
   Step 3: Cancel K on the left side. $\quad \dfrac{K\sqrt{P}}{K} = \dfrac{Q}{K}$
   cancel-↑

   $\sqrt{P} = \dfrac{Q}{K}$

Step 4: Square both sides. $\qquad (\sqrt{P})^2 = (Q/K)^2$

Step 5: The formula is developed. $\qquad P = (Q/K)^2$

7. The formula: $FL = PDP - NP$

    Develop a formula for NP.

    SOLUTION HELP

    Step 1: Transfer $-NP$ to the left side. $\qquad FL + NP = PDP$

    Step 2: Transfer FL to the right side. $\qquad NP = PDP - FL$

    Step 3: The formula is developed. $\qquad NP = PDP - FL$

8. The formula: $FL = CQ^2L$

    Develop a formula for Q.

    SOLUTION HELP

    Step 1: Change the sides of the formula. $\qquad CQ^2L = FL$

    Step 2: Divide both sides by CL. $\qquad \dfrac{CQ^2L}{CL} = \dfrac{FL}{CL}$

    Step 3: Cancel CL on the left side. $\qquad$ cancel-↓---↓ $\dfrac{CQ^2L}{CL} = \dfrac{FL}{CL}$ cancel-↑-↑

    Step 4: Take the square root of each side. $\qquad \sqrt{Q^2} = \sqrt{FL/CL}$

    Step 5: The formula is developed. $\qquad Q = \sqrt{FL/CL}$

9. The formula: $h = 1.13Hg$

    Develop a formula for Hg.

    SOLUTION HELP

    Step 1: Change the sides of formula. $\qquad 1.13\,Hg = h$

    Step 2: Divide both sides by 1.13. $\qquad \dfrac{1.13\,Hg}{1.13} = \dfrac{h}{1.13}$

    Step 3: Cancel the 1.13's on the left side. $\qquad$ cancel-↓ $\dfrac{1.13\,Hg}{1.13} = \dfrac{h}{1.13}$ cancel-↑

    Step 4: The formula is developed. $\qquad Hg = \dfrac{h}{1.13}$

Chapter Summary

This chapter introduced the methods considered important when manipulating formulas. It provided the two important words when working with formulas, then presented sample problems using formulas that contained positive and negative units, formulas concerned with multiplying units, units being squared, and units that contain square root signs. The reader was then given a number of problems to practice the manipulation of formulas.

Section One
U.S. Units of Measurement

In the following tables, any italicized *foot* or *mile* is a survey foot or U.S. statute mile, rather than an international foot or mile.

Units of Length

12 inches (in)	= 1 foot (ft)
3 feet	= 1 yard (yd)
$16\frac{1}{2}$ *feet*	= 1 rod (rd), pole, or perch
40 rods	= 1 furlong (fur) = 660 *feet*
8 furlongs	= 1 U.S. statute mile (mi) = 5,280 *feet*
1,852 meters	= 6,076.11549 feet (approximately)
	= 1 international nautical mile

Units of Area

144 square inches (in²)	= 1 square foot (ft²)
9 square feet	= 1 square yard (yd²)
	= 1,296 square inches
$272\frac{1}{4}$ square *feet*	= 1 square rod (sq rd)
160 square rods	= 1 acre = 43,560 square *feet*
640 acres	= 1 square *mile* (mi²)
1 *mile* square	= 1 section of land
6 *miles* square	= 1 township
	= 36 sections = 36 square *miles*

Units of Volume

1,728 cubic inches (in³)	= 1 cubic foot (ft³)
27 cubic feet	= 1 cubic yard (yd³)

Source: National Institute of Standards and Technology

Units of Liquid Volume[1]

4 gills (gi)	=	1 pint (pt) = 28.875 cubic inches
2 pints	=	1 quart (qt) = 57.75 cubic inches
4 quarts	=	1 gallon (gal) = 231 cubic inches
	=	8 pints = 32 gills

Apothecaries' Units of Liquid Volume

60 minims (min or)	=	1 fluid dram (fl dr or f)
	=	0.225 6 cubic inches
8 fluid drams	=	1 fluid ounce (fl oz or f)
	=	1.804 7 cubic inches
16 fluid ounces	=	1 pint = 28.875 cubic inches
	=	128 fluid drams
2 pints	=	1 quart (qt) = 57.75 cubic inches
	=	32 fluid ounces = 256 fluid drams
4 quarts	=	1 gallon (gal) = 231 cubic inches
	=	128 fluid ounces = 1,024 fluid drams

Units of Dry Volume[2]

2 pints (pt)	=	1 quart (qt) = 67.200 6 cubic inches
8 quarts	=	1 peck (pk) = 537.605 cubic inches
	=	16 pints
4 pecks	=	1 bushel (bu) = 2,150.42 cubic inches
	=	32 quarts

---

[1] When necessary to distinguish the liquid pint or quart from the dry pint or quart, the word "liquid" or the abbreviation "liq" should be used in combination with the name or abbreviation of the liquid unit.

[2] When necessary to distinguish the dry pint or quart from the liquid pint or quart, the word "dry" should be used in combination with the name or abbreviation of the dry unit.

Section Two
Metric Units of Measurement

Units of Length

10 millimeters (mm)	= 1 centimeter (cm)	
10 centimeters	= 1 decimeter (dm)	= 100 millimeters
10 decimeters	= 1 meter (m)	= 1,000 millimeters
10 meters	= 1 dekameter (dam)	
10 dekameters	= 1 hectometer (hm)	= 100 meters
10 hectometers	= 1 kilometer (km)	= 1,000 meters

Units of Area

100 square millimeters (mm^2)	= 1 square centimeter (cm^2)	
100 square centimeters	= 1 square decimeter (dm^2)	
100 square decimeters	= 1 square meter (m^2)	
100 square meters	= 1 square dekameter (dam^2)	= 1 are
100 square decameters	= 1 square hectometer (hm^2)	= 1 hectare (ha)
100 square hectometers	= 1 square kilometer (km^2)	

Units of Liquid Volume

10 milliliters (mL)	= 1 centiliter (cL)	
10 centiliters	= 1 deciliter (dL)	= 100 milliliters
10 deciliters	= 1 liter[3]	= 1,000 milliliters
10 liters	= 1 dekaliter (daL)	
10 dekaliters	= 1 hectoliter (hL)	= 100 liters
10 hectoliters	= 1 kiloliter (kL)	= 1,000 liters

Source: National Institute of Standards and Technology

Units of Volume[3]

1,000 cubic millimeters (mm³)	=	1 cubic centimeter (cm³)
1,000 cubic centimeters	=	1 cubic decimeter (dm³)
	=	1,000,000 cubic millimeters
1,000 cubic decimeters	=	1 cubic meter (m³)
	=	1,000,000 cubic centimeters
	=	1,000,000,000 cubic millimeters

Units of Mass

10 milligrams (mg)	=	1 centigram (cg)
10 centigrams	=	1 decigram (dg) = 100 milligrams
10 decigrams	=	1 gram (g) = 1,000 milligrams
10 grams	=	1 dekagram (dag)
10 dekagrams	=	1 hectogram (hg) = 100 grams
10 hectograms	=	1 kilogram (kg) = 1,000 grams
1,000 kilograms	=	1 megagram (Mg) or 1 metric ton (t)

---

[3]By action of the 12th General Conference on Weights and Measures (1964), the liter is a special name for the cubic decimeter.

Section Three
International Units of Measurement

Units of Length—International Measure
(all underlined figures are exact)

Units		Inches	Feet	Yards	Miles	Centimeters	Meters
1 inch	=	1	0.083 333 33	0.027 777 78	0.000 015 782 83	2.54	0.025 4
1 foot	=	12	1	0.333 333 3	0.000 189 393 9	30.48	0.304 8
1 yard	=	36	3	1	0.000 568 181 8	91.44	0.914 4
1 mile	=	63 360	5 280	1 760	1	160 934.4	1609.344
1 centimeter	=	0.393 700 8	0.032 808 40	0.010 936 13	0.000 006 213 712	1	0.01
1 meter	=	39.370 08	3.280 840	1.093 613	0.000 621 371 2	100	1

Units of Area—International Measure
(all underlined figures are exact)

Units		Square Inches	Square Feet	Square Yards
1 square inch	=	1	0.006 944 444	0.000 771 604 9
1 square foot	=	144	1	0.111 111 1
1 square yard	=	1296	9	1
1 square mile	=	4 014 489 600	27 878 400	3 097 600
1 square centimeter	=	0.155 000 3	0.001 076 391	0.000 119 599 0
1 square meter	=	1550.003	10.763 91	1.195 990

Units		Square Miles	Square Centimeters	Square Meters
1 square inch	=	0.000 000 000 249 097 7	6.451 6	0.000 645 16
1 square foot	=	0.000 000 035 870 06	929.030 4	0.092 903 04
1 square yard	=	0.000 000 322 830 6	8361.273 6	0.836 127 36
1 square mile	=	1	25 899 881 103.36	2 589 988.110 336
1 square centimeter	=	0.000 000 000 038 610 22	1	0.0001 1
1 square meter	=	0.000 000 386 102 2	10 000	1

Source: National Institute of Standards and Technology

Units of Volume
(all underlined figures are exact)

Units	Cubic Inches	Cubic Feet	Cubic Yards
1 cubic inch =	1	0.000 578 703 7	0.000 021 433 47
1 cubic foot =	1728	1	0.037 037 04
1 cubic yard =	46 656	27	1
1 cubic centimeter =	0.061 023 74	0.000 035 314 67	0.000 001 307 951
1 cubic decimeter =	61.023 74	0.035 314 67	0.001 307 951
1 cubic meter =	61 023.74	35.314 67	1.307 951

Units	Milliliters (Cubic Centimeters)	Liters (Cubic Decimeters)	Cubic Meters
1 cubic inch =	16.387 064	0.016 387 064	0.000 016 387 064
1 cubic foot =	28 316.846 592	28.316 846 592	0.028 316 846 592
1 cubic yard =	764 554.857 984	764.554 857 984	0.764 554 857 984
1 cubic centimeter =	1	0.001	0.000 001
1 cubic decimeter =	1000	1	0.001
1 cubic meter =	1 000 000	1000	1

Section Four
Addition Chart

	1	2	3	4	5	6	7	8	9	10
1	2	3	4	5	6	7	8	9	10	11
2	3	4	5	6	7	8	9	10	11	12
3	4	5	6	7	8	9	10	11	12	13
4	5	6	7	8	9	10	11	12	13	14
5	6	7	8	9	10	11	12	13	14	15
6	7	8	9	10	11	12	13	14	15	16
7	8	9	10	11	12	13	14	15	16	17
8	9	10	11	12	13	14	15	16	17	18
9	10	11	12	13	14	15	16	17	18	19
10	11	12	13	14	15	16	17	18	19	20

To use this chart, select a digit in the left-hand vertical column that you wish to add to a digit in the top horizontal column, or select a digit in the top horizontal column that you wish to add to a digit in the left-hand vertical column. Where the two intersect provides the answer. For example, if you wish to add 7 and 6, select the 7 from the left-hand vertical column and the 6 from the top horizontal column, or the 7 from the horizontal column and the 6 from the left-hand vertical column. Note that they intersect at 13. Therefore, $7 + 6 = 13$.

Section Five
Multiplication Chart

	1	2	3	4	5	6	7	8	9
1	1	2	3	4	5	6	7	8	9
2	2	4	6	8	10	12	14	16	18
3	3	6	9	12	15	18	21	24	27
4	4	8	12	16	20	24	28	32	36
5	5	10	15	20	25	30	35	40	45
6	6	12	18	24	30	36	42	48	54
7	7	14	21	28	35	42	49	56	63
8	8	16	24	32	40	48	56	64	72

This chart can be used in one of two ways: A number in the top horizontal column can be multiplied by a number in the left vertical column, or a number in the left vertical column can be multiplied by a number in the top horizontal column. Where the two columns intersect provides the answer. For example, if you want to multiply 7 × 6, select the 7 from the left-hand vertical column and the 6 from the top horizontal column, or the 6 from the left-hand vertical column and the 7 from the top horizontal column. Note that the two columns intersect at 42, the answer to the multiplication.

Section Six
Times Tables

The Two Tables
2 × 2 = 4
2 × 3 = 6
2 × 4 = 8
2 × 5 = 10
2 × 6 = 12
2 × 7 = 14
2 × 8 = 16
2 × 9 = 18

The Three Tables
3 × 2 = 6
3 × 3 = 9
3 × 4 = 12
3 × 5 = 15
3 × 6 = 18
3 × 7 = 21
3 × 8 = 24
3 × 9 = 27

The Four Tables
4 × 2 = 8
4 × 3 = 12
4 × 4 = 16
4 × 5 = 20
4 × 6 = 24
4 × 7 = 28
4 × 8 = 32
4 × 9 = 36

The Five Tables
5 × 2 = 10
5 × 3 = 15
5 × 4 = 20
5 × 5 = 25
5 × 6 = 30
5 × 7 = 35
5 × 8 = 40
5 × 9 = 45

The Six Tables
6 × 2 = 12
6 × 3 = 18
6 × 4 = 24
6 × 5 = 30
6 × 6 = 36
6 × 7 = 42
6 × 8 = 48
6 × 9 = 54

The Seven Tables
7 × 2 = 14
7 × 3 = 21
7 × 4 = 28
7 × 5 = 35
7 × 6 = 42
7 × 7 = 49
7 × 8 = 56
7 × 9 = 63

The Eight Tables
8 × 2 = 16
8 × 3 = 24
8 × 4 = 32
8 × 5 = 40
8 × 6 = 48
8 × 7 = 56
8 × 8 = 64
8 × 9 = 72

The Nine Tables
9 × 2 = 18
9 × 3 = 27
9 × 4 = 36
9 × 5 = 45
9 × 6 = 54
9 × 7 = 63
9 × 8 = 72
9 × 9 = 81

Section Seven
Placement Chart

2	3	4	5	6	.	3	5	7	8
ten thousands place	thousands place	hundreds place	tens place	units place	decimal place	tenths place	hundredths place	thousandths place	ten thousandths place
10,000	1,000	100	10	1		1/10	1/100	1/1,000	1/10,000

Section Eight
Basic Fire Department Hydraulic Figures

1. Water weighs approximately 62.5 pounds per cubic foot.
2. There are approximately 7.48 gallons of water in a cubic foot.
3. One gallon of water weighs approximately 8.35 pounds.
4. There are 1,728 cubic inches in a cubic foot.
5. One gallon of water contains approximately 231 cubic inches.
6. A column of water measuring one inch by one inch at its base and one foot high weighs approximately .434 pounds.
7. A column of water measuring one inch by one inch at its base and 2.304 feet high weighs approximately one pound.

Section Nine
Abbreviations

A (or a) = area or altitude (height)

AFL = amount of friction loss in an appliance

B (or b) = base (also the whole to which the percentage is a part)

BP = back pressure (the pressure in psi that is working against the pumper)

C = friction loss coefficient

D (or d) = diameter

Dis = discharge

F = force

FL (or fl) = friction loss in psi (pounds per square inch)

FP = forward pressure (the pressure in psi that is aiding the pumper when the pumper is supplying water to an elevation below the pumper)

G = gallons

H (or h) = height or head

Hg = inches of mercury

L (or l) = length

NP (or p) = nozzle pressure (the pressure at the nozzle tip in psi)

P = pressure in psi or percentage

Pi = 3.1416

Prepiping = prepiping friction loss

psi = pounds per square inch

Q = amount or quantity of flow

R (or r) = radius or rate (the percent expressed as a decimal)

RPDP = required pumper discharge pressure in psi

S (or s) = side

V = volume of a container or the velocity of flow

W (or w) = width

Section Ten
Traditional Formulas Introduced in the Book

- The required pump discharge pressure can be determined by the formula:

  $$RPDP = NP + FL + BP + AFL + Prepiping - FP$$

  where   RPDP = required pump discharge pressure
  NP = nozzle pressure
  FL = friction loss
  BP = back pressure
  AFL = appliance friction loss
  Prepiping = prepiping friction loss
  FL = forward pressure

- The formula for determining the volume of a rectangular container is:

  $$V = l\, w\, h$$

  where   l = length
  w = width
  h = height

- The force on a surface is determined by the formula:

  $$F = p \times a$$

  where   p = pressure
  a = area

- To determine the number of gallons in a rectangular container, use the formula:

  $$\text{gallons} = 7.48\, V$$

  where   V = volume

- To determine the number of pounds of water in a cylindrical container, use the formula:

  $$\text{pounds} = 62.5\, V$$

  where   V = volume

- To determine the pressure created by a column of water, use the formula:

$$P = .434\, H$$

where H = head

- To determine the head when the pressure at an outlet is known, use the formula:

$$H = 2.304\, P$$

where P = pressure

- The number of gallons of water that a rectangular container can hold is determined by the formula:

$$\text{gallons} = 7.48\, V$$

where V = volume

- Determining the height to which water will rise can be determined by the formula:

$$\text{height} = 1.13\, Hg$$

where Hg = inches of mercury

- To determine the pressure at an outlet, two formulas can be used:

$$p = .434\, h \quad \text{or}$$
$$p = \frac{h}{2.304}$$

where h = head

- The formula for determining the velocity flow is:

$$vf = 12.14\sqrt{p}$$

where p = nozzle pressure

- The formula for determining the nozzle reaction is:

$$NR = 1.57\, D^2 P$$

where D = diameter of tip
       P = nozzle pressure

- The formula for determining the discharge from a nozzle tip is:

$$Dis = 29.7\, D^2 \sqrt{P}$$

where D = diameter of tip
       P = nozzle pressure

- The friction loss in a hose layout is determined by the formula:

$$FL = CQ^2 L$$

where C = friction loss coefficient
       Q = flow in hundreds of gpm (flow/100)
       L = hose length in hundreds of feet (length/100)

- To determine the volume of a cylindrical container, use the formula:

$$V = .7854 D^2 H$$

  where D = diameter of container

  H = height of container

- The formula for finding the area of a square is:

$$\text{area} = S^2$$

  where S = side

- The formula for determining the area of a rectangle is:

$$\text{area} = L \times W$$

  where L = length

  W = width

- The formula for determining the area of a right triangle is:

$$\text{area} = \frac{ab}{2}$$

  where a = height

  b = length

- The National Fire Academy uses a rectangle as a portion of their formula for determining the amount of water required for extinguishing and controlling a structure fire. The formula is:

$$\text{fire flow} = \frac{l \times w}{3}$$

  where l = length

  w = width

- The formula for determining the area of a circle is:

$$a = \text{pi } r^2$$

or

$$a = .7854 D^2$$

  where r = radius

  D = diameter

- Several formulas used in the fire service measure pressure. The most common is:

$$P = .434 H$$

  where H = head

  Another formula is:

$$BP = .434 H$$

  where BP = back pressure

  H = head

Another formula is:

$$FP = .434\,H$$

where   FP = forward pressure
               H = head

- The formula for determining the weight of water in a rectangular or cylindrical container is:

$$\text{weight} = 62.5\,V$$

where   V = volume

- The formula used for determining the amount of water in a hose line is:

$$\text{gallon capacity} = \frac{\text{volume (in cubic inches)}}{231}$$

where   v = volume of hose line

The formula used to convert Fahrenheit temperature to centigrade temperature is:

$$C = \frac{(5)(F - 32)}{9}$$

where   C = centigrade
               F = Fahrenheit

Section Eleven
Square Roots

Square Roots of Numbers (1 to 250)

n	$\sqrt{n}$	n	$\sqrt{n}$	n	$\sqrt{n}$	n	$\sqrt{n}$	n	$\sqrt{n}$	n	$\sqrt{n}$
1	1.	32	5.6569	63	7.9373	94	9.6954	125	11.1803		
2	1.414	33	5.7446	64	8.0000	95	9.7468	126	11.2250		
3	1.732	34	5.8310	65	8.0623	96	9.7980	127	11.2694		
4	2.000	35	5.9161	66	8.1240	97	9.8489	128	11.3137		
5	2.236	36	6.0000	67	8.1854	98	9.8995	129	11.3578		
6	2.449	37	6.0828	68	8.2462	99	9.9449	130	11.4018		
7	2.646	38	6.1644	69	8.3066	100	10.0000	131	11.4455		
8	2.828	39	6.2450	70	8.3666	101	10.0499	132	11.4891		
9	3.000	40	6.3246	71	8.4261	102	10.0995	133	11.5326		
10	3.162	41	6.4031	72	8.4853	103	10.1489	134	11.5758		
11	3.3166	42	6.4807	73	8.5440	104	10.1980	135	11.6190		
12	3.4641	43	6.5574	74	8.6023	105	10.2470	136	11.6619		
13	3.6056	44	6.6332	75	8.6603	106	10.2956	137	11.7047		
14	3.7417	45	6.7082	76	8.7178	107	10.3441	138	11.7473		
15	3.8730	46	6.7823	77	8.7750	108	10.3923	139	11.7898		
16	4.0000	47	6.8557	78	8.8318	109	10.4403	140	11.8322		
17	4.1231	48	6.9282	79	8.8882	110	10.4881	141	11.8743		
18	4.2426	49	7.0000	80	8.9443	111	10.5357	142	11.9164		
19	4.3589	50	7.0711	81	9.0000	112	10.5830	143	11.9583		
20	4.4721	51	7.1414	82	9.0554	113	10.6301	144	12.0000		
21	4.5826	52	7.2111	83	9.1104	114	10.6771	145	12.0416		
22	4.6904	53	7.2801	84	9.1652	115	10.7238	146	12.0830		
23	4.7958	54	7.3485	85	9.2195	116	10.7703	147	12.1244		
24	4.8990	55	7.4163	86	9.2736	117	10.8167	148	12.1655		
25	5.0000	56	7.4833	87	9.3274	118	10.8628	149	12.2066		
26	5.0990	57	7.5498	88	9.3808	119	10.9087	150	12.2474		
27	5.1962	58	7.6158	89	9.4340	120	10.9545	151	12.2882		
28	5.2915	59	7.6811	90	9.4868	121	11.0000	152	12.3288		
29	5.3852	60	7.7460	91	9.5394	122	11.0454	153	12.3693		
30	5.4772	61	7.8102	92	9.5917	123	11.0905	154	12.4097		
31	5.5678	62	7.8740	93	9.6437	124	11.1355	155	12.4499		

156	12.4900	175	13.2288	194	13.9284	213	14.5945	232	15.2315
157	12.5300	176	13.2665	195	13.9642	214	14.6287	233	15.2643
158	12.5698	177	13.3041	196	14.0000	215	14.6629	234	15.2971
159	12.6095	178	13.3417	197	14.0357	216	14.6969	235	15.3297
160	12.6491	179	13.3791	198	14.0712	217	14.7309	236	15.3623
161	12.6886	180	13.4164	199	14.1067	218	14.7648	237	15.3948
162	12.7279	181	13.4536	200	14.1421	219	14.7986	238	15.4272
163	12.7671	182	13.4907	201	14.1774	220	14.8324	239	15.4596
164	12.8062	183	13.5287	202	14.2127	221	14.8661	240	15.4919
165	12.8452	184	13.5647	203	14.2478	222	14.8997	241	15.5242
166	12.8841	185	13.6015	204	14.2829	223	14.9332	242	15.5563
167	12.9228	186	13.6882	205	14.3178	224	14.9666	243	15.5885
168	12.9615	187	13.6748	206	14.3527	225	15.0000	244	15.6205
169	13.0000	188	13.7113	207	14.3875	226	15.0333	245	15.6525
170	13.0384	189	13.7477	208	14.4222	227	15.0665	246	15.6844
171	13.0767	190	13.7840	209	14.4568	228	15.0997	247	15.7162
172	13.1149	191	13.8203	210	14.4914	229	15.1327	248	15.7480
173	13.1529	192	13.8564	211	14.5258	230	15.1658	249	15.7797
174	13.1909	193	13.8924	212	14.5602	231	15.1987	250	15.8114

Glossary

abstract Not related to a concrete object.

addend A number or quantity to be added to another number or quantity.

adding the digits A system for determining the excess of nines of a number.

area measurement A two-dimensional measurement that measures the surface of an area.

arithmetic The science or art of computation by positive, real numbers.

back pressure The pressure working back against the pump in the apparatus when the water is being supplied at a level above the pumper outlet.

body heat The degree of heat in a living organism.

braces A grouping symbol indicated by { }. The use of braces is limited.

brackets A grouping symbol indicated by [ ].

capacity Uses the ounce as the basic unit of measurement.

casting out nines method A system for determining the excess of nines of a number.

centi One-hundredth.

central tendency Means of measuring or describing the distribution of numerical values within a set of data.

circle A two-dimensional geometric figure that is formed by a center point, with every point on the formed line being an equal distance from the center point.

composite number A whole number is either a prime number or a composite number. A composite number is a whole number derived from multiplying together two or more prime numbers.

data The plural of datum.

datum Refers to a single piece of information.

deca 10.

deci One-tenth.

decimal number A number made up of a whole number and a decimal fraction.

denominator The bottom number in a fraction.

difference The answer to the problem, which is the difference between the minuend and the subtrahend.

digit One of the Arabic numbers, 0 to 9, used to represent numbers in the customary system.

dividend The number to be divided.

divisor The number that will be used to divide the dividend into equal parts.

equation A mathematical statement that two expressions divided by an equal sign are of the same value.

equivalent fractions A fraction that represents the same amount as another fraction, although the two fractions have different numbers.

"excess of nines" A number from 1 up to and including 8 that remains in a two-digit or larger number after all nines, or combination of numbers totaling nine, are removed from the number.

exponent A symbol to indicate how many times a number is to be multiplied by itself. It is placed to the right and raised up above the number that is to be acted upon.

flow pressure The pressure taken after the water has been placed in motion.

fluid A substance that flows freely.

foot (ft) 12 inches.

force The amount of energy being applied at a given point, or points.

formula An equation designed to obtain the answer to a mathematical problem.

forward pressure A situation where pressure aid is given to a water source such as a pump when the water being used is below the water source.

fractions Used to represent different parts of a whole number or object.

friction loss The amount of pressure lost in a hose line due to water moving through the line.

graph A visual one- or two-dimensional figure that shows the relationship between two or more variables.

head The vertical distance from the surface of the water to the point being considered.

heat index The "feels like," or apparent temperature.

hecto 100.

humidity The amount of moisture in the air.

hurricanes Tropical depressions that originate in areas of low pressure in the equatorial regions of the Atlantic or Caribbean.

improper fraction A fraction with a numerator equal to or larger than the denominator.

inch The basic unit of length for the customary system.

integers Whole numbers that have no fractional or decimal parts.

kilo 1,000.

length Uses the inch as the basic unit of measurement.

like fractions Two or more fractions that have the same denominator.

line graphs Generally compare the relationship of one variable to another.

mass Uses an ounce as the basic unit of measurement.

mean The mathematic average value of the data in a set.

median The middle value in a distribution when the values within the distribution have been arranged according to their magnitudes.

mentally Carried out in the mind without the use of any physical aid.

meter The basic unit for length measurement in the metric system.

mile (mi) 5,280 feet.

milli One-thousandth.

minuend The upper number of a subtraction problem.

minuend fraction The fraction from which the subtrahend fraction is to be subtracted.

mixed number A whole number and a fraction, with the fraction having a value of less than 1.

mode The value that most frequently appears in a distribution.

multiplicand The number being multiplied.

multiplier The number of times the multiplicand is multiplied.

negative integers Numbers such as −1, −2, −3, and so on.

numerator The top number in a fraction.

parentheses A grouping symbol indicated by ( ).

percent A portion of a whole.

percentage A portion of a whole that does not have an identification symbol.

percentile A value on a scale of 100 that indicates the percent of a distribution that is equal to or below it.

period A group of three digits used for counting purposes.

pie charts Generally used when it is desired to show the percentage parts of a whole.

positive integers Generally referred to as the counting numbers.

pound The standard unit of measurement for weight in the customary system.

pressure Force per unit area.

prime numbers Numbers greater than 1 that can be divided only by themselves.

product The answer to the multiplication.

proper fraction A fraction with a numerator smaller than denominator.

quotient The answer to the division problem.

radical symbol A symbol placed over a number to indicate that the square root of the number is to be found. The symbol is $\sqrt{}$.

real numbers Includes whole numbers, integers, fractions, and decimal fractions.

rectangle A two-dimensional geometric figure having four right (90°) angles.

relative humidity The amount of moisture in the air compared with the amount of moisture the air can hold at that temperature.

repeating decimal A decimal whose fraction repeats itself forever.

required pump discharge pressure The total pressure that a pumper must provide in order to take care of all the pressure losses in a layout plus the pressure required at the nozzle.

residual pressure The pressure remaining at a hydrant outlet once the water is flowing.

right rectangle An oblong plane surface with parallel sides and four right (90°) angles.

rounding a number Converting a number to an approximation of that number.

skewed When the true meaning or nature of something is misrepresented.

square A geometric figure having four right (90°) angles and four equal sides.

squaring a number Multiplying a number by itself.

static pressure The pressure of water when it is not in motion.

statistics A branch of mathematics that is concerned with the collection, analysis, interpretation, and presentation of numerical data.

subtrahend The lower number of a subtraction problem.

subtrahend fraction The fraction to be subtracted.

sum The result obtained by adding together two or more digits or numbers.

symbol A written or printed mark representing something abstract.

temperature A degree of heat.

terminating decimal A decimal whose decimal fraction stops.

tornado An extremely destructive funnel-shaped rotating column of air that travels in a narrow path over land.

triangle A two-dimensional geometric figure having three sides and three angles.

unlike fractions Two or more fractions that have different denominators.

velocity flow How fast water is moving through hose or out a nozzle or other opening.

volume The size of a three-dimensional space that is enclosed within an object or occupied by an object.

weight The downward force created by an object due to the earth's gravity in the customary system.

word problem A problem where it is up to the reader to arrange the problem in the proper form for solving it.

yard (yd) 3 feet.

zero The integer 0 is neither positive nor negative.

Index

Abbreviations, 269
Abstract, 2
Added to, 26
Addend, 13
Adding the digits, 18
Addition (adding)
   chart, 264
   decimals, 128–130
   fractions, 95–100
   mixed numbers, 98–100
   negative and positive integers, 23–25
   order of operations, 152, 156
   proving, 17–21, 129–130
   review of, 14–17
   word problems, 25–28
Apothecaries' units of liquid volume, 259
Area measurements
   of a circle, 9, 11, 122, 180, 272
   converting, 178
   customary units of, 258
   defined, 178
   formulas, 272
   international units of, 262
   metric units of, 260
   problems using, 180–182
   of a rectangle, 117, 148, 179, 272
   of a square, 178, 272
   of a triangle, 179–180, 272
Arithmetic, 2–3

Back pressure (BP), 29, 72
   determining, 73, 184, 272
   formulas, 272
Bar graphs, 230
Blood pressure, 182
Body heat, 207
Braces, 152
Brackets, 152
   order of operations, 152–153

Canceling, 110–112
Casting out nines, 18

Centi, 168
Centigrade (Celsius), converting from
   Fahrenheit to, 207–209
Central tendency, 224, 225
Charts
   addition, 264
   defined, 228
   development of, 229
   multiplication, 265
   pie, 232–233
   placement, 267
   problems using, 233–240
   types of, 229
Circle, area of, 9, 11, 122, 180
Combination problems
   adding and subtracting, 152, 156
   exponents and square roots, 152, 153–154
   multiplying and dividing, 152, 154–156
   order of operations, 152
   parentheses and brackets, 152–153
   symbols used in, 152
Combined total, 26
Common fractions
   adding, 95–100
   dividing, 118–120
   multiplying, 109–114
   subtracting, 104–106
   types of, 88–89
Composite numbers, 2, 5, 89
Cross-multiplication, 91
Customary system, 4, 168
   See also Measurements
   fluid measurements, 202–207
   pressure measurements, 182–191
   temperature measurements, 207–209
   units of measurement, 258–259
Customary system, area measurements
   of a circle, 9, 11, 122, 180
   converting, 178
   defined, 178
   problems using, 180–182
   of a rectangle, 117, 148, 179

Customary system, area measurements (*Continued*)
  of a square, 178
  of a triangle, 179–180
  units of, 258
Customary system, length measurements
  converting, 170–171
  description of, 170, 171–174
  mixed numbers, setting problem up correctly, 171–174
  problems using, 174–178
  units of, 258
Customary system, volume measurements
  converting, 192
  of cylindrical containers, 117–118, 158, 164, 193–194
  defined, 191
  problems using, 195–198
  of rectangular containers, 10, 11, 54, 116, 142–143, 158, 192–193
  units of, 258
Customary system, weight measurements
  converting, 199
  defined, 198
  determining weight of water in a container, 163, 199–200
  problems using, 200–202
Cylindrical container
  determining pounds, 140, 270
  determining volume, 117–118, 158, 164, 193–194, 272
  formulas for, 270, 272

Data, 224
Datum, 224
Deca, 168
Deci, 168
Decimal fractions, parts of, 124
Decimal number, 124
Decimals, 88
  adding, 128–130
  applications, 123–124
  converting, to percent, 213
  converting fractions to, 126
  converting percent to, 212
  converting to fractions, 91, 126–127, 212–213
  dividing, 145–146
  multiplying, 138–139
  repeating, 125
  rounding numbers, 125–126
  subtracting, 134–135
  terminating, 125
Denominator
  defined, 88
  finding a common, 95
Diastolic pressure, 182
Difference, 68, 104
Digit, 2
Dividend, 59, 145
Divisibility rules, 5–6
Division (dividing)
  answer, proving, 61
  decimals, 145–146

  determining if dividend can be divided evenly, 62–63
  four-digit dividend by a two-digit divisor, 61–62
  fractions, 118–120
  mixed numbers by mixed numbers, 119–120
  a number by 5, 64
  a number by 10, 64
  a number by 25, 64
  a number by 50, 64
  order of operations, 152, 154–156
  positive and negative integers, 65
  review of, 60–61
  shortcuts, 64
  whole numbers by mixed numbers, 119
Divisor, 59, 145
Dry volume, units of, 259

Egyptian number system, 4
Equations, defined, 241
Equivalent fractions
  changing a fraction to, 90
  defined, 88
Equivalent nozzle pressure formula, 86
Excess of nines method, 17, 18–21, 35–36, 129–130
Exponents, 77
  in formulas, 246–247
  order of operations, 152, 153–154

Fahrenheit, converting from Centigrade (Celsius) to, 207–209, 273
Fever, 207
FLOW, 30, 72
Flow pressure, 183
Fluid
  defined, 202
  measurements, 202–207
  problems using, 204–207
Force
  defined, 183
  determining, 54–55
  formula, 270
Formulas
  defined, 241
  double trouble, 248–252
  exponents in, 246–247
  manipulating, 242
  multiplication units, 244–246
  positive and negative units, 242–244
  problems, solving, 241–242
  problems using, 253–257
  square roots in, 247–248
  used in this book, 270–273
Forward pressure (FP), 184–185, 273
Fractions
  adding, 95–100
  changing to equivalent, 90
  changing to higher or lower term, 90–91
  converting, to decimals, 126
  converting, to percent, 214–215

converting decimals to, 91, 126–127, 212–213
converting percent to, 213
converting improper fractions to mixed numbers, 91
converting minutes to, 92
converting mixed numbers to improper, 91
defined, 88
dividing, 118–120
equivalent, 88, 90
improper, 88, 89
like, 89
minuend, 104
mixed numbers, 88, 89
multiplying, 109–114
proper, 88
subtracting, 104–106
subtrahend, 104
types of common, 88–89
unlike, 89
whole numbers and, 89–90
Friction loss (FL), 29, 72
coefficients for single hose lines, 185
determining, 10–11, 75, 76, 162–163, 185–186
formulas for, 252, 271

Grams, 199
Graphs
bar, 230
defined, 225, 228
development of, 229
line, 230–231
parts of, 228–229
problems using, 233–240
types of, 229
Grouping symbols, 152
order of operations, 152–153

Head, 183, 271
Hecto, 168
Horizontal axis, 228
Horizontal reach, determining, 84, 251
Hose line
determining amount of, 32, 33–34, 36
determining gallon capacity of, 158–159
Hydraulic figures, basic, 268
Hydraulic formulas and problems, 10–11, 157–165

Improper fractions
converting, to mixed numbers, 91
converting mixed numbers to, 91
defined, 88, 89
Integers
See also Negative integers; Positive integers
defined, 2, 4, 22
International units of measurement, 262–263

Kilo, 168

Length measurements
converting, 170–171

customary, 170, 171–174
customary units of, 258
international units of, 262
metric, 169–171
metric units of, 260
mixed numbers, setting problem up correctly, 171–174
problems using, 174–178
unit symbols, how to use, 170–171
Lift, determining, 143
Like fractions, 89
Line graphs, 230–231
Liquid volume
apothecaries' units of, 259
customary units of, 259
metric units of, 260
Liter, 192

Mass, metric units of, 261
Mean, 225–226
Measurements
See also Customary system; Metric system
area, 178–182
changing prefixes, 168–169
customary, 4, 168
customary units of, 258–259
fluid, 202–207
international units of, 262–263
length, 169–178
metric, 4, 168
metric units of, 260–261
pressure, 182–191
temperature, 207–209
terminology, 167
time, 169
volume, 191–198
weight, 198–202
Median, 224, 226–227
Mentally, 2
Meter (metre)
equivalencies, 169–170
prefixes with, 169
Metric system, 4, 168
See also Measurements
fluid measurements, 202–207
pressure measurements, 182–191
temperature measurements, 207–209
units of measurement, 260–261
Metric system, area measurements
of a circle, 9, 11, 122, 180
converting, 178
defined, 178
problems using, 180–182
of a rectangle, 117, 148, 179
of a square, 178
of a triangle, 179–180
units of, 260
Metric system, length measurements
converting, 170–171

Metric system, length measurements (*Continued*)
   description of, 169–171
   problems using, 174–178
   units of, 260
   unit symbols, how to use, 170–171
Metric system, volume measurements
   converting, 192
   of cylindrical containers, 117–118, 158, 164, 193–194
   defined, 191
   problems using, 195–198
   of rectangular containers, 10, 11, 54, 116, 142–143, 158, 192–193
   units of, 260–261
Metric system, weight measurements
   converting, 199
   defined, 198
   determining weight of water in a container, 163, 199–200
   problems using, 200–202
Milli, 168
Minuend, 68, 104
Minuend fraction, 104
Minutes, converting to fractions, 92
Mixed numbers
   adding, 98–100
   converting, to improper fractions, 91
   converting improper fractions to, 91
   defined, 88, 89
   dividing mixed numbers by mixed numbers, 119–120
   dividing whole numbers by, 119
   multiplying whole numbers, common fractions and, 113–114
   subtracting, 104–106
Mode, 224, 227
Multiplicand, 38
Multiplication (multiplying)
   canceling, 110–112
   chart, 265
   decimals, 138–139
   fractions, 109–114
   mixed numbers, whole numbers, and common fractions, 113–114
   mixing whole digits, 44
   a number by 100, 50
   a number by 10, 49–50
   one-digit, 44
   order of operations, 152, 154–156
   of positive and negative integers, 51–52
   problems, proving, 45–46
   problems, solving, 42–44
   review of, 39
   shortcut procedures, 46–51
   single-digit number by 9, 40–42
   squaring a number ending in 5, 49
   symbol and the use of the word "of," 109
   tables, 39–40, 266
   two-digit, 44
   two-digit number by eleven, 47
   two teen numbers, 47–49
   whole numbers and common fractions, 112–113
Multiplier, 38

Negative integers
   adding, 23–25
   defined, 2, 4, 7–8, 22
   dividing, 65
   multiplying, 51–52
   subtracting, 70–71
Nozzle pressure (NP), 29, 72
   determining square root of, 85
   equivalent formula, 86
Nozzle reaction (NR)
   determining, 9–10, 11, 118, 160–161
   formulas for, 251, 271
   momentary, 122
Nozzle tip discharge
   determining, 30, 54, 55, 56, 58, 144, 161–162
   formulas for, 249–250, 251, 252, 271
Number(s)
   basic concepts, 2–3
   composite, 2, 5, 89
   explanation of, 4–5
   mixed, 88, 89
   systems, 3–4
   whole numbers and fractions, 89–90
Numerator, 88

Order of operations
   adding and subtracting, 152, 156
   defined, 152
   exponents and square roots, 152, 153–154
   how to remember, 152
   multiplying and dividing, 152, 154–156
   order of operations, 152
   parentheses and brackets, 152–153
   symbols used in, 152
Origin, 228

Parentheses, 152
   order of operations, 152–153
Percent
   converting, to a decimal, 212
   converting, to a fraction, 213
   converting decimals to, 213
   converting fractions to, 214–215
   defined, 211, 212
   problems, solving, 215–218
   problems using, 220–222
   symbol for, 212
Percentages, 88
   defined, 211, 218
   problems using, 220–222
   sale items examples, 219–220
   tipping examples, 218–219
Percentile, 218
Per centum, 212
Period, 2, 5

Pie charts, 232–233
Placement chart, 267
Positive integers
  adding, 23–25
  defined, 2, 4, 7–8, 22
  dividing, 65
  multiplying, 51–52
  subtracting, 70–71
Pound, 198
Pressure
  blood, 182
  defined, 182, 183
  determining, 140–142, 144, 147–148
  formulas, 183–184, 271, 272–273
  measurements, 182–191
  problems using, 187–191
  water, 183–187
Prime numbers
  defined, 2, 5
  determining, 5–7
  fractions and, 95
Product, 38
Proper fractions, 88

Quotient, 59, 145

Radical symbol, 77
Ratio
  defined, 211
  expressions of, 212
Real numbers, 2, 4
Rectangle and rectangular containers
  area of, 117, 148, 179, 272
  cubic feet, 147
  cubic inches, 54, 56, 58
  formulas for, 251, 270, 271
  gallons of water in, 140, 142, 270, 271
  volume, 10, 11, 54, 116, 142–143, 158,
    192–193, 251, 270
Repeating decimals, 125
Required pump discharge pressure (RPDP), 72
  determining, 30, 31–32, 34–35, 36–37, 102–103,
    132, 164–165, 186–187
  formula for, 251–252, 270
Residual pressure, 183
Right triangles, formulas for, 159–160
Rounding numbers, 125–126

Sale items examples, 219–220
Skewed, 225
Square, area of, 178, 272
Square roots
  in formulas, 247–248
  order of operations, 152, 153–154
  problem, proving, 82–83
  review of, 77–80
  table, 274–275
  two-digit problem, 80–81
Squaring a number ending in five, 49

Standard curve of distribution, 227–228
Static pressure, 183
Statistics
  defined, 224, 225
  graphs and charts, 228–233
  mean, 225–226
  median, 224, 226–227
  mode, 224, 227
  problems using, 233–240
  standard curve of distribution, 227–228
  terminology, 224–225
Subtraction (subtracting)
  decimals, 134–135
  fractions, 104–106
  mixed numbers, 104–106
  order of operations, 152, 156
  positive and negative integers, 70–71
  problem, proving, 69–70
  review of, 68–69
Subtrahend, 68, 104
Subtrahend fraction, 104
Sum, 13
Sum of, 26
Symbols
  defined, 2
  grouping, 152
  percent, 212
  radical, 77
Systolic pressure, 182

Temperature measurements, 207–209
Terminating decimals, 125
Time, measurements of, 169
Tipping examples, 218–219
Total of, 26
Triangle, area of, 179–180, 272

Unlike fractions, 89

Velocity flow
  determining, 84–85, 118, 160, 251
  formulas, 251, 271
Vertical axis, 228
Vertical reach, determining, 85
Volume
  apothecaries' units of liquid, 259
  converting, 192
  customary units of, 258–259
  of cylindrical containers, 117–118, 158, 164, 193–194
  defined, 191
  formulas, 251, 270, 272
  international units of measurement, 263
  measurements, 191–198
  metric units of, 261
  problems using, 195–198
  of rectangular containers, 10, 11, 54, 116, 142–143, 158,
    192–193, 251, 270, 272
  units of dry, 259
  units of liquid, 259

Water
- determining amount dropped from an aircraft, 54, 76, 77, 165
- determining amount in a container, 163, 199–200, 270
- determining amount needed to extinguish a fire, 272
- formulas, 270, 271, 272, 273
- gallon capacity of, 118, 270, 271
- height when drafting, 122, 271
- pressure, 183–187
- weight capacity of, 118, 157, 273

Weight
- converting, 199
- defined, 198
- determining weight of water in a container, 163, 199–200, 273
- measurements, 198–202
- problems using, 200–202

Whole numbers
- dividing, by mixed numbers, 119
- fractions and, 89–90
- multiplying common fractions and, 112–113
- multiplying mixed numbers, common fractions and, 113–114

Word problems
- addition, 25–28
- changing, to number problems, 9
- how to solve, 8–9
- sample, 9–11

Zero, 2, 5